SUSTAINABLE HYDRAULICS IN THE ERA OF GLOBAL CHANGE

Sustainable Hydraulics in the Era of Global Change

Advances in Water Engineering and Research

Editors

Sébastien Erpicum, Benjamin Dewals,
Pierre Archambeau & Michel Pirotton
*Hydraulics in Environmental and Civil Engineering (HECE),
Department ArGEnCo, University of Liege (ULg), Liege, Belgium*

CRC Press
Taylor & Francis Group
Boca Raton London New York Leiden

CRC Press is an imprint of the
Taylor & Francis Group, an **informa** business

A BALKEMA BOOK

Cover photos:
Front cover photo: Lanaye locks, situated across the Belgian-Dutch border in the north of Liege, are a key connexion in the European navigation network. The 4th lock of Lanaye (225 m × 25 m, 14 m head difference), together with its pumping station and hydropower plant (2.3 MW), was awarded the 1st IAHR Industry Innovation Award.

Courtesy: Parafly – http://www.parafly.be/
Service Public de Wallonie – http://voies-hydrauliques.wallonie.be/

CRC Press/Balkema is an imprint of the Taylor & Francis Group, an informa business

© 2016 Taylor & Francis Group, London, UK

Typeset by MPS Limited, Chennai, India
Printed and bound in Great Britain by CPI Group (UK) Ltd, Croydon, CR0 4YY

All rights reserved. No part of this publication or the information contained herein may be reproduced, stored in a retrieval system, or transmitted in any form or by any means, electronic, mechanical, by photocopying, recording or otherwise, without written prior permission from the publishers.

Although all care is taken to ensure integrity and the quality of this publication and the information herein, no responsibility is assumed by the publishers nor the author for any damage to the property or persons as a result of operation or use of this publication and/or the information contained herein.

Published by: CRC Press/Balkema
P.O. Box 11320, 2301 EH Leiden, The Netherlands
e-mail: Pub.NL@taylorandfrancis.com
www.crcpress.com – www.taylorandfrancis.com

ISBN: 978-1-138-02977-4 (Set of Hbk and CDROM)
ISBN: 978-1-4987-8149-7 (eBook PDF)

Table of contents

Preface XV
Committees XVII
Support from institutions and scientific societies XIX
Sponsors XXI

Keynotes

Co-development of coastal flood models: Making the leap from expert analysis to decision support 3
B.F. Sanders, A. Luke, J.E. Schubert, H.R. Moftakhari, A. AghaKouchak, R.A. Matthew, K. Goodrich, W. Cheung, D.L. Feldman, V. Basolo, D. Houston, K. Serrano, D. Boudreau & A. Eguiarte

Challenges and opportunities for research and technological innovation in the water sector throughout Europe 4
P. Balabanis

Hydro-environment and eco-hydraulics

Tsunami seismic generation and propagation: Validity of the shallow water equations 7
M. Le Gal & D. Violeau

Free surface flow through homogeneous bottom roughness 8
H. Romdhane, A. Soualmia, L. Cassan & D. Dartus

Study of turbulent flow through large porous media 9
M. Jouini, A. Soualmia, G. Debenest & L. Masbernat

Vorticity fluxes on the wake of cylinders within random arrays 10
A.M. Ricardo & R.M.L. Ferreira

Effect of downstream channel slope on numerical modelling of dam break induced flows 11
A.E. Dinçer, Z. Bozkuş & A.N. Şahin

Analysis of dotation discharge impact at a fishway entrance via numerical 3D CFD simulation 12
J. Klein & M. Oertel

Characterization of flow resistance in a floodplain for varying building density 13
S. Guillén-Ludeña, D. Lopez, E. Mignot & N. Rivière

Turbulent dispersion in bounded horizontal jets. RANS capabilities and physical modeling comparison 14
D. Valero, D.B. Bung & M. Oertel

Modelling methane emissions from Vilar reservoir (Portugal) 15
R.F. Lino, P.A. Diogo, A.C. Rodrigues & P.S. Coelho

The impact of stormwater overflows on stream water quality 16
T. Julínek & J. Říha

Groundwater quality of Feriana-Skhirat in central Tunisia and its sustainability for agriculture and drinking purposes 17
I. Hassen, F. Hamzaoui-Azaza & R. Bouhlila

Modelling the transport and decay of microbial tracers in a macro-tidal estuary 18
A.A. Bakar, R. Ahmadian & R.A. Falconer

Building water and chemicals budgets over a complex hydrographic network 19
V. Carbonnel, N. Brion, M. Elskens, P. Claeys & M.A. Verbanck

Evaluating the ecological restoration of a Mediterranean reservoir 20
P. Sidiropoulos, M. Chamoglou & I. Kagkalou

Spatio-temporal interaction of morphological variability with hydrodynamic parameters
in the scope of integrative measures at the river Danube 21
M. Glas, M. Tritthart, M. Liedermann, P. Gmeiner & H. Habersack

Analysis of contributions and uncertainties of fish population models for the development
of river continuity concepts in the river basin Ruhr, Germany 22
D. Teschlade & A. Niemann

Hydrologic and hydraulic design to reduce diffuse pollution from drained peatlands 23
S. Mohammadighavam & B. Kløve

Flow modelling and investigation of flood scenarios on the Cavaillon River, Haiti 24
A. Joseph, N. Gonomy, Y. Zech & S. Soares-Frazão

Simulating runoff generation and consumption for rehabilitation of downstream ecosystems
in arid northwest China 25
C. He & L. Zhang

Integrating Spatial Multi Criteria Decision Making (SMCDM) with Geographic Information
Systems (GIS) for determining the most suitable areas for artificial groundwater recharge 26
M. Zare & M. Koch

Hydrological simulation in the Tiete basin 27
T. Rocha, I.G. Hidalgo, A.F. Angelis & J.E.G. Lopes

Hydraulic analysis of an irrigation headworks complex in the Artibonite department (Haïti) 28
S. Louis, N. Gonomy, N. de Ville & M.A. Verbanck

Effect of emergent vegetation distribution on energy loss 29
E. Eriş, G. Bombar & Ü. Kavaklı

Monthly reservoir operating rules generated by implicit stochastic optimization and
self-organizing maps 30
C.A.S. Farias, E.C.M. Machado & L.N. Brasiliano

Comparison of PIV measurements and CFD simulations of the velocity field over bottom racks 31
L.G. Castillo, J.T. García, J.M. Carrillo & A. Vigueras-Rodríguez

Image processing techniques for velocity estimation in highly aerated flows: Bubble Image
Velocimetry vs. Optical Flow 32
D.B. Bung & D. Valero

Biomechanical tests of aquatic plant stems: Techniques and methodology 33
A.M. Łoboda, Ł. Przyborowski, M. Karpiński & R.J. Bialik

LS-PIV procedure applied to a plunging water jet issuing from an overflow nappe 34
Y. Bercovitz, F. Lebert, M. Jodeau, C. Buvat, D. Violeau, L. Pelaprat & A. Hajczak

PIV-PLIF characterization of density currents 35
B. Pérez-Díaz, P. Palomar & S. Castanedo

Physical model tests for the construction stages of large breakwaters. Case studies to the
ports of Barcelon and Cruña (Spain) 36
R.M. Gutierrez, J. Lozano & J.M. Valdés

Low-cost 3D mapping of turbulent flow surfaces 37
A. Nichols & M. Rubinato

Sediment transport measurements in the Schelde-estuary: How do acoustic backscatter, optical
transmission and direct sampling compare? 38
S. Thant, Y. Plancke & S. Claeys

Field-deployable particle image velocimetry with a consumer-grade digital camera applicable
for shallow flows 39
K. Koca, A. Lorke & C. Noss

Numerical modelling of meandering jets in shallow rectangular reservoir using two different turbulent closures — 40
Y. Peltier, S. Erpicum, P. Archambeau, M. Pirotton & B. Dewals

Turbulent momentum exchange over a natural gravel bed — 41
C. Miozzo, A. Marion, A. Nichols & S.J. Tait

Water as a renewable energy

Planning a small hydro-power plant in Lübeck (Germany) – Who owns the water? — 45
M. Oertel

Effect of changes in flow velocity on the phytobenthic biofilm below a small scale low head hydropower scheme — 46
L. O'Keefe & S. Ilic

New concepts in small hydropower plants schemes in Romania — 47
F. Popa, D. Florescu & B. Popa

Fish behavioral and mortality study at intake and turbine — 48
P. Rutschmann, S. Schäfer, B. Rutschmann & F. Geiger

Long-term evaluation of the wave climate and energy potential in the Mediterranean Sea — 49
G. Lavidas, V. Venugopal & A. Agarwal

Numerical simulation and validation of CECO wave energy converter — 50
P. Rosa-Santos, M. López-Gallego, F. Taveira-Pinto, D. Perdigão & J. Pinho-Ribeiro

Pumped hydroelectric energy storage: A comparison of turbomachinery configurations — 51
A. Morabito, J. Steimes & P. Hendrick

Cost and revenue breakdown for a pumped hydroelectric energy storage installation in Belgium — 52
J. Steimes, G. Al Zohbi, P. Hendrick, B. Haut & S. Doucement

Numerical investigations of a water vortex hydropower plant implemented as a fish ladder – Part I: The water vortex — 53
N. Lichtenberg, O. Cleynen & D. Thévenin

Performance mapping of ducted free-stream hydropower devices — 54
O. Cleynen, S. Hoerner & D. Thévenin

In-situ scale testing of current energy converters in the Sea Scheldt, Flanders, Belgium — 55
T. Goormans, S. Smets, R.J. Rijke, J. Vanderveken, J. Ellison & R. Notelé

Coupling of water expansion and production of energy on public water distribution network — 56
B. Michaux & C. Piro

Coastal aspects in the era of global change

Impact study of a new marina on the sediment balance. Numerical simulation — 59
M.C. Khellaf & H.E. Touhami

Numerical simulation of scour in front of a breakwater using OpenFoam — 60
N. Karagiannis, T. Karambas & C. Koutitas

Application of an unstructured grid tidal model in the Belgian Continental Shelf — 61
B.B. Sishah, M. Chen & O. Gourgue

Online coupling of SWAN and SWASH for nearshore applications — 62
M. Ventroni, A. Balzano & M. Zijlema

Spectral parameters modification in front of a seawall with wave return — 63
Th. Giantsi, G. Papacharalampous, N. Martzikos & C.I. Moutzouris

Renewal time scales in tidal basins: Climbing the Tower of Babel — 64
D.P. Viero & A. Defina

Morphodynamic processes at Mariakerke: Numerical model implementation — 65
R. Silva, R. De Sutter & R. Delgado

Design features of the upcoming coastal and ocean basin in Ostend, Belgium 66
*A. Kortenhaus, P. Troch, N. Silin, V. Nelko, P. Devriese, V. Stratigaki, J. De Maeyer,
J. Monbaliu, E. Toorman, P. Rauwoens, D. Vanneste, T. Suzuki, T. Van Oyen & T. Verwaest*

Idealized model study on tidal wave propagation in prismatic and converging basins with tidal flats 67
T. Boelens, T. De Mulder, H. Schuttelaars & G. Schramkowski

Safety standards for the coastal dunes in The Netherlands 68
R.P. Nicolai, M. Kok, A.W. Minns, S. van Vuren & C.J.J. Eijgenraam

Impact of the sea level rise on low lying areas of coastal zone: The case of Batu Pahat 69
A.B. Yannie, A.H. Radzi, A. Dunstan & W.H.M. Wan Mohtar

A database of validation cases for tsunami numerical modelling 70
*D. Violeau, R. Ata, M. Benoit, A. Joly, S. Abadie, L. Clous, M. Martin Medina, D. Morichon,
J. Chicheportiche, M. Le Gal, A. Gailler, H. Hébert, D. Imbert, M. Kazolea, M. Ricchiuto,
S. Le Roy, R. Pedreros, M. Rousseau, K. Pons, R. Marcer, C. Journeau & R. Silva Jacinto*

Wave force calculations due to wave run-up on buildings: A comparison of formulas applied
in a real case 71
F.G. Brehin, N. Zimmermann, V. Gruwez & A. Bolle

Coastal tsunami-hazard mapping 72
C. Yavuz & E. Kentel

Waves generated by ship convoy: Comparison of physical and numerical modeling with
in-situ measurements 73
T. Cohen Liechti, G. De Cesare, A.J. Schleiss, R. Amacher & M. Pfister

Example of wave impact on a residential house 74
D. Wüthrich, M. Pfister & A.J. Schleiss

Innovation and sustainability in hydraulic engineering and water resources

Estimation of rainfall-runoff relation using HEC-HMS for a basin in Turkey 77
H. Akay, M. Baduna Koçyiğit & A.M. Yanmaz

Using ANN and ANFIS Models for simulating and predicting Groundwater Level Fluctuations
in the Miandarband Plain, Iran 78
M. Zare & M. Koch

A numerical groundwater flow model of Bursa Basköy aquifer 79
S. Korkmaz & G.E. Türkkan

Groundwater management and potential climate change impacts on Oum Er Rbia basin, Morocco 80
M. El Azhari & D. Loudyi

Flow modeling in vegetated rivers 81
M. Mabrouka

Comparison of methods to calculate the shear velocity in unsteady flows 82
G. Bombar

Non-linear optimization of a 1-D shallow water model and integration into Simulink for
operational use 83
L. Goffin, B. Dewals, S. Erpicum, M. Pirotton & P. Archambeau

Limitation of self-organization within a confined aquifer 84
M.C. Westhoff, S. Erpicum, P. Archambeau, M. Pirotton, B. Dewals & E. Zehe

Preliminary experiments on the evolution of river dunes 85
G. Oliveto & M.C. Marino

A conceptual sediment transport simulator based on the particle size distribution 86
B.T. Woldegiorgis, W. Bauwens, M. Chen, F. Pereira & A. van Griensven

Effect of seepage on the friction factor in an alluvial channel 87
P. Mahesh, V. Deshpande & B. Kumar

Experimental investigation of light particles transport in a tidal bore generated in a flume — 88
L. Thomas, W. Reichl, Y. Devaux, A. Beaudouin & L. David

Effect of hydrodynamics factors on flocculation processes in estuaries — 89
A. Mhashhash, B. Bockelmann-Evans & S. Pan

Lack of scale separation in granular flows driven by gravity — 90
G. Rossi & A. Armanini

Sediment transport in the Schelde-estuary: A comparison between measurements, transport formula and numerical models — 91
Y. Plancke & G. Vos

Influence of non-uniform flow conditions on riverbed stability: The case of smooth-to-rough transitions — 92
D. Duma, S. Erpicum, P. Archambeau, M. Pirotton & B. Dewals

The probabilistic solution of dike breaching due to overtopping — 93
Z. Alhasan, D. Duchan & J. Říha

Field measurements and numerical modelling on local scour around a ferry slip structure — 94
P. Penchev, V. Bojkov & V. Penchev

Experimental and numerical study of scour downstream Toachi Dam — 95
L.G. Castillo, M. Castro, J.M. Carrillo, D. Hermosa, X. Hidalgo & P. Ortega

Design of scour protections and structural reliability techniques — 96
T. Fazeres-Ferradosa, F. Taveira-Pinto, L. das Neves & M.T. Reis

Failure of fluvial dikes: How does the flow in the main channel influence the breach development? — 97
I. Rifai, S. Erpicum, P. Archambeau, D. Violeau, M. Pirotton, B. Dewals & K. El kadi Abderrezzak

Continuous grid monitoring to support sediment management techniques — 98
T. Van Hoestenberghe, R. Vanthillo, P. Heidinger, J. Dornstädter, N. Dezillie & N. Van Ransbeeck

River flow analysis with adjoints – An efficient, universal methodology to quantify spatial interactions and sensitivities — 99
U.H. Merkel, J. Riehme & U. Naumann

Comparative use of FVM and integral approach for computation of water flow in a coiled pipe and a surge tank — 100
M. Torlak, B. Šeta & A. Bubalo

Seepage characteristics in embankments subject to variable water storages on both sides — 101
M. Calamak & A.M. Yanmaz

Characterization of nappe vibration on a weir — 102
M. Lodomez, M. Pirotton, B. Dewals, P. Archambeau & S. Erpicum

Shock wave patterns in supercritical junction manholes with inlet bottom offsets — 103
G. Crispino, C. Gisonni & M. Pfister

Investigation of the hydrodynamic pressures on lock gates during earthquakes — 104
L. Buldgen, P. Rigo & H. Le Sourne

Stilling basin design for inlet sluice with vertical drop structure: Scale model results vs. literature formulae — 105
J. Vercruysse, K. Verelst & T. De Mulder

Supercritical flow around an emerged obstacle: Hydraulic jump or wall-jet-like bow-wave? — 106
G. Vouaillat, N. Rivière, G. Launay & E. Mignot

Monsin movable dam in Belgium: A case study — 107
C. Swartenbroekx, C. Savary & D. Bousmar

Numerical modelling of contracted sharp crested weirs — 108
A. Duru, A.B. Altan-Sakarya & M.A. Kokpinar

Field measurements at the new lock of Lanaye (Belgium) before the opening to navigation — 109
C. Savary, D. Bousmar, C. Swartenbroekx, G. Zorzan & T. Auguste

Experimental investigation of the influence of breaking logs on the flow patterns induced
by lock filling with gate openings 110
K. Verelst, J. Vercruysse, P.X. Ramos & T. De Mulder

Relation between free surface profiles and pressure profiles with respective fluctuations
in hydraulic jumps 111
J.D. Nóbrega, H.E. Schulz & M.G. Marques

Experimental study of head loss through an angled fish protection system 112
H. Boettcher, R. Gabl, S. Ritsch & M. Aufleger

Scale effects for air-entraining vortices at pipe intake structures 113
K. Taştan

Analysis of various Piano Key Weir geometries concerning discharge coefficient development 114
M. Oertel & F. Bremer

Determination of discharge coefficient of triangular labyrinth side weirs with one and two
cycles using the nonlinear PLS method 115
M.A. Nekooie, A. Parvaneh & A. Kabiri-Samani

Discharge capacity of conventional side weirs in supercritical conditions 116
A. Parvaneh, M.R. Jalili Ghazizadeh, A. Kabiri-Samani & M.A. Nekooie

Discharge coefficient of oblique labyrinth side weir 117
A. Kabiri-Samani, A. Parvaneh & M.A. Nekooie

Reconstruction of a stage-discharge relation for a damaged weir on the Cavaillon river, Haïti 118
O. Carlier d'Odeigne, O. Roelandts, T. Verschoore, Y. Zech & S. Soares-Frazão

A two-fluid SPH model for landslides 119
A. Ghaïtanellis, D. Violeau, A. Leroy, A. Joly & M. Ferrand

Hydraulic modelling strategies for flood mapping. Application to coastal area in central Vietnam 120
V. Ngoc Duong & P. Gourbesville

Estimation of 1D-confluence model parameters in right-angled discordant beds' confluences using
3D numerical model 121
D. Đorđević & I. Stojnić

Practical application of numerical modelling to overbank flows in a compound river channel 122
P.M. Moreta, A.J. Rotimi, A.S. Kawuwa & S. López-Querol

Comparison between different methods to compute the numerical fluctuations in path-conservative
schemes for SWE-Exner model 123
F. Carraro, V. Caleffi & A. Valiani

A new Osher Riemann solver for shallow water flow over fixed or mobile bed 124
D. Zugliani & G. Rosatti

Grid coarsening and uncertainty in 2D hydrodynamic modelling 125
V. Bellos & G. Tsakiris

Migration characteristics of a meandering river: The Madhumati river, Bangladesh 126
M.S. Banda & S. Egashira

Employing surrogate modelling for the calibration of a 2D flood simulation model 127
V. Christelis, V. Bellos & G. Tsakiris

Estimating stem-scale mixing coefficients in low velocity flows 128
F. Sonnenwald, I. Guymer, A. Marchant, N. Wilson, M. Golzar & V. Stovin

SLIM: A model for the land-sea continuum and beyond 129
*E. Deleersnijder, S. Blaise, P. Delandmeter, T. Fichefet, E. Hanert, J. Lambrechts, Y. Le Bars,
V. Legat, J. Naithani, C. Pham Van, J.-F. Remacle, S. Soares-Frazão, C. Thomas, V. Vallaeys,
D. Vincent, T. Hoitink, M. Sassi, V. Dehant, O. Karatekin & E. Wolanski*

Hydrometeorological extremes, uncertainties and global change

Hybrid downscaling and conditioning for characterizing multivariate flooding extremes *M. del Jesus, P. Camus & I.J. Losada*	133
Application of artificial neural networks in meteorological drought forecasting using Standard Precipitation Index (SPI) *S. Golian, P. Yavari & H. Ruigar*	134
Accuracy assessment of ISI-MIP and FAO hydrological modelling results in the Upper Indus Basin *A. Khan & M. Khan*	135
A Gaussian design-storm for Mediterranean convective events *I. Andrés-Doménech, R. García-Bartual, M. Rico Cortés & E. Albentosa Hernández*	136
Extreme hydrological situations on Danube River – Case study Bezdan hydrological station (Serbia) *M. Urošev, I. Leščešen, D. Štrbac & D. Dolinaj*	137
Presenting an empirical model for determining the sugar beet evapotranspiration by GDD parameter (Case study: Torbat-Jam, Iran) *F. Jan Nesar, A. Khashei Suiki, S.R. Hashemi & S. Moradi Kashkooli*	138
Impact of rainfall variability on the sewerage system of Casablanca city, Morocco *L. Ennajem & D. Loudyi*	139
Implications of CMIP5 derived climate scenarios for discharge extremes of the Rhine *F.C. Sperna Weiland, M. Hegnauer, H. Van den Boogaard, H. Buiteveld, R. Lammersen & J. Beersma*	140
How will be future rainfall IDF curves in the context of climate change? *H. Tabari, P. Hosseinzadehtalaei, P. Willems, S. Saeed, E. Brisson & N. Van Lipzig*	141
Influence of model uncertainty on real-time flood control performance *E. Vermuyten, P. Meert, V. Wolfs & P. Willems*	142
Flash flood prediction, case study: Oman *E. Holzbecher, A. Al-Qurashi, F. Maude & M. Paredes-Morales*	143
Equipping the TRENT2D model with a WebGIS infrastructure: A smart tool for hazard management in mountain regions *N. Zorzi, G. Rosatti, D. Zugliani, A. Rizzi & S. Piffer*	144
Reservoir operation applying a discrete hedging rule with ensemble streamflow prediction to cope with droughts *S. Lee & Y. Jin*	145
Determination of the interaction between surface flow and drainage discharge *S. Kemper, A. Schlenkhoff & J. Balmes*	146
Effects of flow orientation on the onset of motion of flooded vehicles *C. Arrighi, F. Castelli & H. Oumeraci*	147
Development of a computationally efficient urban flood modelling approach *V. Wolfs, V. Ntegeka, D. Murla Tuyls & P. Willems*	148
Rapid flood inundation modelling in a coastal urban area using a surrogate model of the 2D shallow water equations *L. Cea, M. Bermudez, J. Puertas, I. Fraga & S. Coquerez*	149
Impacts of urban expansion on future flood damage: A case study in the River Meuse basin, Belgium *A. Mustafa, M. Bruwier, J. Teller, P. Archambeau, S. Erpicum, M. Pirotton & B. Dewals*	150
Uncertainty assessment of river water levels on energy head loss through hydraulic control structures *P. Meert, F. Pereira & P. Willems*	151
Floods and cultural heritage: Risk assessment and management for the city of Florence, Italy *C. Arrighi, F. Castelli & B. Mazzanti*	152
Assessing the impact of climate change on extreme flows across Great Britain *L. Collet, L. Beevers & C. Prudhomme*	153

Continuous hydrologic modeling of coastal plain watershed using HEC-HMS 154
M. Pourreza-Bilondi & S. Zahra Samadi

A methodology to account for rainfall uncertainty at the event scale in fully distributed
rainfall runoff models 155
I. Fraga, L. Cea, J. Puertas, M. Álvarez-Enjo, S. Salsón & A. Petazzi

Estimating probability of dike failure by means of a Monte Carlo approach 156
R. Van Looveren, T. Goormans, J. Blanckaert, K. Verelst & P. Peeters

Developing a 100-point scoring system for quality assessment of ADCP streamflow measurements 157
H. Huang

An alternate approach for assessing impacts of climate change on water resources: Combining
hazard likelihood and catchment sensitivity 158
B. Grelier, G. Drogue, M. Pirotton, P. Archambeau & B. Dewals

Efficient management of inland navigation reaches equipped with lift pumps in a
climate change context 159
H. Nouasse, A. Doniec, E. Duviella & K. Chuquet

Evaluation of changes storm precipitations during century for the modeling of floods 160
V.V. Ilinich & T.D. Larina

Main impacts of climate change on seaport construction and operation 161
B. Koppe

Special session: Innovative solutions for adaptation of European hydropower systems in view of climate and market changes

Experimental assessment of head losses through elliptical and sharp-edged orifices 165
N.J. Adam, G. De Cesare & A.J. Schleiss

Integrated assessment of underground pumped-storage facilities using existing coal mine infrastructure 166
R. Alvarado Montero, T. Wortberg, J. Binias & A. Niemann

Feasibility assessment of micro-hydropower for energy recovery in the water supply network
of the city of Fribourg 167
I. Samora, P. Manso, M.J. Franca, A.J. Schleiss & H.M. Ramos

Banki-Michell micro-turbines for energy production in water distribution networks 168
V. Sammartano, P. Filianoti, G. Morreale, M. Sinagra & T. Tucciarelli

How groundwater interactions can influence UPSH (Underground Pumping Storage
Hydroelectricity) operations 169
S. Bodeux, E. Pujades, P. Orban, S. Brouyère & A. Dassargues

Integration of hydropower plant within an existing weir – "A hidden treasure" 170
M. Marence, J.S. Ingabire & B. Taks

Special session: Buoyancy-driven flows

Numerical simulation of periodically forced convective currents in aquatic canopies 173
M. Tsakiri & P. Prinos

Experiments on the impact of snow avalanches into water 174
G. Zitti, C. Ancey, M. Postacchini & M. Brocchini

Two dimensional Lattice Boltzmann numerical simulation of a buoyant jet 175
A. Montessori, P. Prestininzi, M. La Rocca, D. Malcangio & M. Mossa

Interfacial instabilities of gravity currents in the presence of surface waves 176
L.M. Stancanelli, R.E. Musumeci & E. Foti

Unraveling salt fluxes: A tool to determine flux components and dispersion rates from 3D models 177
W.M. Kranenburg, T. van der Kaaij, H. van den Boogaard, R.E. Uittenbogaard & Y.M. Dijkstra

Density currents flowing up a slope 178
L. Ottolenghi, C. Adduce, R. Inghilesi, V. Armenio & F. Roman

Turbulent entrainment in a gravity current 179
M. Holzner, M. van Reeuwijk & H. Jonker

Remote sensing and coastal morphodynamic modelling: A review of current approaches
and future perspectives 180
M. Benincasa, F. Falcini, C. Adduce & R. Santoleri

A comparison of simple buoyant jet models with CFD analysis of overflow dredging plumes 181
B. Decrop, T. De Mulder & E. Toorman

Special session: Management of hydraulic systems by means of fuzzy logic

Assessment of annual hydrological drought based on fuzzy estimators 185
M. Spiliotis, P. Angelidis & B. Papadopoulos

Fuzzy regression analysis between sediment transport rates and stream discharge in the case
of two basins in northeastern Greece 186
M. Spiliotis, V. Kitsikoudis & V. Hrissanthou

Rainfall data regression model using fuzzy set logic 187
C. Tzimopoulos, C. Evangelides & B. Papadopoulos

Optimal spatial allocation of groundwater under fuzzy hydraulic parameters 188
E. Sidiropoulos

Fuzzy logic uses on eutrophication and water quality predictions 189
G.K. Ellina & I. Kagkalou

Author index 191

Sustainable Hydraulics in the Era of Global Change – Erpicum et al. (Eds.)
© 2016 Taylor & Francis Group, London, ISBN 978-1-138-02977-4

Preface

This book is the Proceedings for the 4th European Congress of the International Association of Hydro-environment engineering and Research (IAHR), which gathered in Liege over 200 hydro-environment researchers and practitioners from all over Europe and beyond.

The overarching theme of the conference is entitled "Sustainable Hydraulics in the Era of Global Change", consistently with the pressing need to design and operate our water systems according to innovative standards in terms of climate adaptation, resource efficiency, sustainability and resilience. This grand challenge triggers unprecedented questions for hydro-environment research and engineering, particularly in relation with our increasingly urbanized societies. Addressing these issues requires a deep understanding of basic processes in fluid mechanics, heat transfer, surface and groundwater flow, among others. These are all themes widely covered by the book, which unveils latest research achievements and innovations relying on state-of-the-art modelling technologies and supported by the exponentially growing availability of data and computation power.

The 4th IAHR Europe Congress was organised by the research group *Hydraulics in Environmental and Civil Engineering* (HECE) of the University of Liege (ULg), supported by several key institutional and commercial sponsors, which are listed in the following pages. The Editors acknowledge the Authors and the members of the conference Committees, who were committed to ensuring a high scientific quality of the papers.

We are confident that the book will serve as a reference for professionals and decision-makers involved in various water-related sectors, such as hydraulic engineering, fluvial hydraulics, coastal engineering, water resources management and many more.

Benjamin DEWALS
Chair of the Organizing Committee

Sébastien ERPICUM
Chair of the International Scientific Committee

May 2016

Committees

ORGANISING COMMITTEE

Chair

Benjamin DEWALS, *University of Liege (ULg), Liege, Belgium*

Members

Pierre ARCHAMBEAU, *University of Liege (ULg), Liege, Belgium*
Daniel BUNG, *Aachen University of Applied Sciences, Aachen, Germany*
Laurence DEFRÈRE, *University of Liege (ULg), Liege, Belgium*
Nadia ELGARA, *University of Liege (ULg), Liege, Belgium*
Sébastien ERPICUM, *University of Liege (ULg), Liege, Belgium*
Patrick HENDRICK, *Université Libre de Bruxelles, Brussels, Belgium*
Michel PIROTTON, *University of Liege (ULg), Liege, Belgium*
Holger SCHÜTTRUMPF, *RWTH Aachen University, Aachen, Germany*
Peter TROCH, *Ghent University, Ghent, Belgium*

IAHR EUROPE DIVISION LEADERSHIP TEAM

Chairs

Anton SCHLEISS (Chair), *Ecole Polytechnique Fédérale de Lausanne (EPFL), Lausanne, Switzerland*
Corrado GISONNI (Vice Chair), 2nd *University of Naples, Naples, Italy*
Aronne ARMANINI (Past Chair), *Universita degli Studi di Trento, Trento, Italy*

Members

Silke WIEPRECHT, *University of Stuttgart, Germany*
Alessandra CROSATO, *UNESCO-IHE Institute for Water Education, the Netherlands*
Francisco TAVEIRA PINTO, *Universidade Do Porto, Portugal*
Manuel GÓMEZ VALENTÍN, *Universidad Politécnica de Cataluña – EHMA, Spain*
Bettina BOCKELMANN-EVANS, *Cardiff School of Engineering, United Kingdom*
Jean-Paul CHABARD, *Electricité de France (EDF), France*
Benjamin DEWALS, *University of Liege (ULg), Belgium*
Markus AUFLEGER, *Innsbruck University, Austria*
Peter RUTSCHMANN, *Technische Universität München, Germany*
Pawel M. ROWINSKI, *Polish Academy of Sciences, Poland*
Ian GUYMER, *University of Warwick, United Kingdom*
Mark RIEDER, *T.G. Masaryk Water Research Institute, Czech Republic*

INTERNATIONAL SCIENTIFIC COMMITTEE

Chair

Sébastien ERPICUM, *University of Liege (ULg), Liege, Belgium*

Members

Jochen ABERLE, *Norway*
Claudia ADDUCE, *Italy*
Pierre ARCHAMBEAU, *Belgium*
Markus AUFLEGER, *Germany*
Francesco BALLIO, *Italy*
Koen BLANCKAERT, *Switzerland*
Bettina BOCKELMANN-EVANS, *UK*
Didier BOUSMAR, *Belgium*
Daniel BUNG, *Germany*
Benoît CAMENEN, *France*
Jean-Paul CHABARD, *France*
Hubert CHANSON, *Australia*
Alessandra CROSATO, *The Netherlands*
Christophe CUDENNEC, *France*
Alain DASSARGUES, *Belgium*
Rita de CARVALHO, *Portugal*
Tom DE MULDER, *Belgium*
Benjamin DEWALS, *Belgium*
Andreas DITTRICH, *Germany*
Dejana DJORDJEVIC, *Serbia*
Rui FERREIRA, *Portugal*
Corrado GISONNI, *Italy*
Manuel GOMEZ VALENTIN, *Spain*
Philippe GOURBESVILLE, *France*
Massimo GRECO, *Italy*
Ian GUYMER, *United Kingdom*
Patrick HENDRICK, *Belgium*
Vlassios HRISSANTHOU, *Greece*
Juha JARVELA, *Finland*
Katinka KOLL, *Germany*
Elpida KOLOKYTHA, *Greece*
Jorge MATOS, *Portugal*
Pedro MANSO, *Switzerland*
Jaak MONBALIU, *Belgium*

Michele MOSSA, *Italy*
Vladimir NIKORA, *United Kingdom*
Mario OERTEL, *Germany*
Stefano PAGLIARA, *Italy*
André PAQUIER, *France*
Gareth PENDER, *United Kingdom*
Fernando PEREIRA, *Belgium*
Michael PFISTER, *Switzerland*
Francisco PINTO, *Portugal*
Michel PIROTTON, *Belgium*
Sebastien PROUST, *France*
Jaromir RIHA, *Czech Republic*
Nicolas RIVIERE, *France*
Pawel ROWINSKI, *Poland*
Nils RUTHER, *Norway*
Paul SAMUELS, *United Kingdom*
Hubert SAVENIJE, *The Netherlands*
Anton SCHLEISS, *Switzerland*
Holger SCHUTTRUMPF, *Germany*
Koji SHIONO, *United Kingdom*
Sandra SOAREZ-FRAZAO, *Belgium*
Thorsten STOESSER, *United Kingdom*
Mutlu SUMER, *Denmark*
James SUTHERLAND, *United Kingdom*
Simon TAIT, *United Kingdom*
Erik TOORMAN, *Belgium*
Peter TROCH, *Belgium*
Blake TULLIS, *United States*
Wim UIJTTEWAAL, *The Netherlands*
Pierre-Louis VIOLLET, *France*
Patrick WILLEMS, *Belgium*
Weiming WU, *United States*
Gerald ZENZ, *Austria*

Sustainable Hydraulics in the Era of Global Change – Erpicum et al. (Eds.)
© 2016 Taylor & Francis Group, London, ISBN 978-1-138-02977-4

Support from institutions and scientific societies

The Congress is sponsored by the *International Association for Hydro-Environment Engineering and Research* (IAHR) and co-sponsored by:

- the *International Association of Hydrological Sciences* (IAHS),
- and *Société Hydrotechnique de France* (SHF).

The Congress is also supported by

- the *University of Liege* (ULg),
- the Belgian research agency *Fund for Scientific Research* (FRS-FNRS).

Sustainable Hydraulics in the Era of Global Change – Erpicum et al. (Eds.)
© 2016 Taylor & Francis Group, London, ISBN 978-1-138-02977-4

Sponsors

MAIN SPONSORS

CO-SPONSORS

Keynotes

Co-development of coastal flood models: Making the leap from expert analysis to decision support

B.F. Sanders, A. Luke, J.E. Schubert, H.R. Moftakhari & A. AghaKouchak
Department of Civil and Environmental Engineering, University of California, Irvine, USA

R.A. Matthew, K. Goodrich, W. Cheung, D.L. Feldman, V. Basolo & D. Houston
Department of Planning, Policy and Design, University of California, Irvine, USA

K. Serrano
Sustainability Initiative, University of California, Irvine, USA

D. Boudreau & A. Eguiarte
Tijuana River National Estuarine Research Reserve, USA

ABSTRACT

Metric resolution "urban" flood models are emerging as powerful tools for analyzing and communicating flood risk at fine spatial and temporal scales which align with personal awareness of geographical areas, and differentiate across individual assets vulnerable to flooding such as homes, businesses, industrial facilities, health care facilities, schools, parks, places of worship, and environmental resources. A combination of trends such as urbanization, intensification of the hydrologic cycle, and higher sea levels portends a significant increase in urban flooding hazards (Hanson et al. 2011). Furthermore, development pressures in many communities mean greater willingness to build in high hazard areas such as floodplains. Urban flood models have potential to provide valuable information about the impacts of development decisions and flood mitigation measures on flood risk, and to support planning for and responding to severe flooding events. However, there is a dearth of knowledge regarding how to transform dense spatiotemporal flood model output data into information that can be used by decision makers. To develop new and improved flood modeling systems, engineers also need to deepen understanding of flooding as a coupled human-water system (Sivapalan et al. 2012).

The Flood Resilient Infrastructure and Sustainable Environments (FloodRISE) project funded by the US National Science Foundation (#1331611) has resulted in the co-development of metric resolution flood models (e.g., Gallien et al. 2011, 2014) in three communities: Newport Beach, Calif.; Tijuana River Valley, Calif.; and Los Laureles in Tijuana, Baja Calif. Co-development of flood models refers to a two-way communication and development process involving the research team and personnel working and living in the study areas, including residents, government officials, emergency managers, civil society groups and business leaders. Activities include in-depth interviews to gather qualitative data about flooding, a formal field survey to gather information about community awareness of and preparedness for flooding, and focus groups to deepen understanding of the decision-points that can be served by flood models and the map formats that best serve decision-making needs. Examples include maps of the 100-year flood depth, the annual probability of flooding, flood intensity (i.e., depth times velocity), and future vs. present flood risk.

Focus groups revealed strong preferences for visualized flood risk information based on decision-making needs. Officials responsible for city planning and regulatory compliance were most interested in the 100-year flood zone presumably as a consequence of flood insurance policy in the USA. Water rescue personnel were most interested flooding intensity and valued greater awareness of the geographical scope of swift water hazards. Civil society groups were most interested the frequency and duration of flooding to assess potential flood damage to local ecosystems. Collectively, results suggest that a diverse set of decision-support needs can be met by a single, fine-scale modeling tool when model output is transformed into a set of maps that communicate different aspects of the flood, and address decision points relevant to diverse users. Further, we conjecture that the process of co-developing the flood model with stakeholders builds confidence in the model among both experts and community members and also creates a solid foundation to plan and evaluate flood risk interventions.

REFERENCES

Gallien, T.W., Sanders, B.F. and Flick R.E. 2014. Urban coastal flood prediction: Integrating wave overtopping, flood defenses, and drainage. Coastal Engineering, 91, 18–28.

Gallien, T.W., Schubert, J.E. and Sanders B.F. 2011. Predicting tidal flooding in urbanized embayments: A modeling framework and data requirements. Coastal Engineering, 58(6), 567–577.

Hanson, S., Nicholls, R., Ranger, N., Hallegatte, S., Corfee-Morlot, J., Herweijer, C., Chateau, J. 2011. A global ranking of port cities with high exposure to climate extremes Climatic Change, 104(1): 89–111.

Sivapalan, M., Savenije, H. H. G., & Blöschl, G. 2012. Sociohydrology: A new science of people and water. *Hydrological Processes*, 26(8), 1270–1276.

Challenges and opportunities for research and technological innovation in the water sector throughout Europe

P. Balabanis
European Commission, Directorate-General for Research & Innovation, Belgium

ABSTRACT

In today's world where intensive use of the world's resources puts pressure on our planet and threatens economic prosperity, growth and jobs, innovative solutions are needed to help us using our resources more efficiently and anticipate more complex demands. Global change, population increase, urbanization are particularly challenging the sustainability of water resources. For the last three consecutive years, the Global Risk report of the World Economic Forum puts water systematically as one of the highest risks that could undermine economic growth.

Citizens, societies, agriculture and industries will increasingly need innovative solutions to meet the need of using water in a more efficient and effective way. Innovative thinking and smarter use of innovation have the potential to bring new solutions quickly and efficiently to the market while responding to the needs of end users in urban, rural and industrial areas. Innovative solutions to water related challenges can directly support wider environmental objectives such as protecting our natural capital and ecosystems, and the biodiversity that supports these. In addition, solutions with regard to drinking water and waste water treatment are to the benefit of public health, which in turn will generate significant savings. Furthermore, solutions to improve protection of, and in, flood-prone areas will enhance public safety and prevent potential economic losses.

Water has been an important activity in successive European Research and Development Framework Programmes over the last decades. Horizon 2020, the current 2014–2020 European Union funding programme for research and innovation, expands the scope of previous water research activities, by addressing the whole chain of research and innovations with the aim of unlocking the innovation potential in the field of water management.

In line with Horizon 2020 objectives, a dedicated focus area "Water innovation: boosting its value for Europe" has been identified in the first Horizon 2020 work programme (2014–2015). This focus area addressed demonstration and market replication activities for eco-innovative, integrated and cross-sectoral solutions for water management. In the second Horizon 2020 work programme actions to boost water innovation for Europe and beyond will be addressed in the areas of the circular economy, sustainable cities, climate services, territorial resilience etc., as well as in other parts of Horizon 2020. Actions strengthening the role of water in the circular economy will be particularly promoted. In this context large scale demonstration/pilot projects, exhibiting a sufficient level of novelty and progress with respect to the state of the art and aiming to implement and test new technological and non-technological solutions through first-of-a-kind experimental development under real solutions, are foreseen. These project should also act as a way to attract most interest from innovators and innovation users (e.g. industries, financial actors, academia, research, private or public entities, regions, cities, citizens and their organizations, etc.), thus helping to unlock additional public/private investments in the water sector and strengthen complementary or synergies with other relevant EU funding mechanisms and initiates, especially, the European Innovation Partnership and Joint Programming Initiative on Water.

REFERENCES

European Commsion, 2011. Recommendation on the research joint programming initiative Water challenges for a changing world, Brussels, OJ C 317, p. 1–3.
European Commission, 2012. European Innovation Partnership water, Strategic Implementation Plan, Brussels.
European Commission, 2013. Regulation of the European Parliament and European Council of 11 December 2013 establishing Horizon 2020 – the Framework Programme for Research and Innovation (2014-2020), Brussels, OJ L 347, p. 81–1030.
World Economic Forum, 2016. The Global Risks Report 2016. 11th Edition, Geneva.

Hydro-environment and eco-hydraulics

Tsunami seismic generation and propagation: Validity of the shallow water equations

M. Le Gal
Saint-Venant Hydraulics Laboratory, University Paris-Est, Chatou, France

D. Violeau
EDF R & D, Chatou, France

ABSTRACT

In this study, we build a linear solution of the free surface for tsunami generation, depending of the two temporal parameters of the seafloor motion: t_r the rise time and v_p the rupture velocity, we use and generalize the work of Hammack (1973) and Todorovska & Trifunac (2001). Our solution is compared to a numerical result and the validity of the Shallow Water Equations (SWE) is studied. The energy ratio ε is defined as the ratio between the potential energy lost under the SWE assumptions (i.e. long waves) over the total potential energy of the tsunami:

$$\varepsilon(t^*) = \frac{E_l(t^*)}{E_t(t^*)} = \frac{\int_{|k^*|\geq 0.2}(\tilde{\eta}^*)^2 dk^*}{\int_{-\infty}^{\infty}(\tilde{\eta}^*)^2 dk^*}.$$

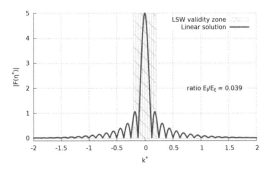

Figure 1. The full blue line represents modulus of $\tilde{\eta}^*$ for $v_p^* = 50$ and $\tau^* = 0$ at the end of ground motion. The gray striped zone shows the domain of validity of the long wave approximation.

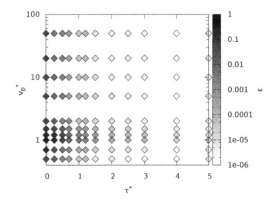

Figure 2. Energy ratio of the lost energy E_l over the total energy E_t, $\varepsilon = \frac{E_l}{E_t}$, at the end of the ground motion as a function of v_p^* and τ^*.

The potential energy is estimated from the Fourier transform of the free surface. The latter is shown on Figure 1 for typical parameters. The energy out of the gray zone is considered as the lost energy. As shown on Figure 2, ε can reach values bigger than 0.5, which means than more than half of the potential energy is lost at the end of the deformation if the SWE are used. Also ε is analysed as a function of time.

REFERENCES

Hammack, J. L. (1973). A note on tsunamis: their generation and propagation in an ocean of uniform depth. *J. Fluid Mech.* 60, 769–799.

Todorovska, M. I. & Trifunac, M. D. (2001). Generation of tsunamis by a slowly spreading uplift of the sea floor. *Soil Dynamics and Earthquake Engineering* 21, 151–167.

Free surface flow through homogeneous bottom roughness

H. Romdhane & A. Soualmia
National Agronomic Institute of Tunisia, University of Carthage, Tunisia

L. Cassan & D. Dartus
Institute of Fluid Mechanics of Toulouse, France

ABSTRACT

The flow in channels and rivers is strongly influenced by the roughness of the funds, by the presence of obstacles, and vegetation.

In this context, an experimental study was conducted in the laboratory of the Institute of Fluid Mechanics of Toulouse "IMFT". The experimental design is a rectangular channel 4 m long and 0.4 m wide and 0.8 m deep. The bottom has a homogeneous roughness contrast (installation of a mat). For the measurement of speed, the channel is equipped with a fast camera.

The originality of our work lies in the application of a PTV particle tracking technique (Particle Tracking Velocimetry). It is a non-intrusive measurement technique to measure instantaneous speed two-dimensional fields in stationary and unsteady flows. It involves seeding the flow through reflective particles, with the same density as the fluid, which will be lit using a lighting plan.

Figure 1. The Channel and these annexes.

Analysis of the results is by processing images taken by the camera using Quick Matlab that is suitable for our case study. The obtained results show a depression of the maximum speed below the free surface. This behavior indicates a delay of the flow in the vicinity of the free surface and this is a direct consequence of the presence of secondary flows in these areas.

These measures are conducted subsequently to the determination of wall parameters.

Study of turbulent flow through large porous media

M. Jouini
National Agronomic Institute of Tunis, University of Carthage, Tunisia

A. Soualmia
Sciences and Technology Water Laboratory (LSTE), Tunisia

G. Debenest
Fluids Mechanics Institute of Toulouse, Group of Porous Mediums (GEMP), France

L. Masbernat
Fluids Mechanics Institute of Toulouse, France

ABSTRACT

In New Caledonia, the mine tailings are protected from flooding by ensuring the water flow through riprap. this study refers to high velocity flow in rockfill. It has been defined by the international firm of engineering consultants MECATER under contract with the Nickel Company of ERAMET group in New Caledonia (SLN), to be carried at the National Agronomic Institute of Tunis (INA Tunis) in collaboration with the Fluids Mechanics Institute of Toulouse (IMF Toulouse). A literature review showed that Darcy relationship is no longer valid for this type of flow (Soualmia et al. 2015; Cyprien 2012). Other researchers like Ergun, Forchheimer, Barree and Conway... have developed models to calculate the pressure drop concerning the inertial forces by integrating a second term on V^2 in Darcy relationship (Barree & Conway 2004; Sano & Kuroiwa 2009). Each relation depends on physical parameters of the porous medium such as permeability, porosity, particle shape.... But it has been shown that the permeability of the studied medium is the most important parameter, since it is the most affected by the flow regime (Wang et al. 1998; Yi et al. 2013). The permeability depends so on Reynolds number. In the frame of this study, many experiments in different types of porous media and under different hydrodynamics conditions were conducted at IMFT. It is observed that the porous media presented a variation of the permeability in different flow regimes. The main aim of this study is to restore the permeability variation curve as a function of the Reynolds number. A comparative analysis between Forchheimer (1901), Ergun (1952) and Barree and Conway (2004) models was presented.

Vorticity fluxes on the wake of cylinders within random arrays

A.M. Ricardo & R.M.L. Ferreira
CEris, Instituto Superior Técnico, ULisboa, Portugal

ABSTRACT

The flow around an isolated cylinder has been extensively studied and flow around multiple cylinders in squared or staggered configurations have also received considerable attention. But the bridge between the knowledge for few cylinders in staggered configuration and the global understanding aimed to a random array is still a challenge as configurations with random distribution of cylinders are less studied.

The interaction between the high background turbulence characteristic of random arrays and vortexes shed by cylinders within arrays affects on how the wake signature of a cylinder spreads, confining the wake and enhancing vorticity cancellation. The mechanism by which the shed vorticity is distributed into the near wake as a result of instabilities in the separating shear layers is not well known, mainly due to the lack of resolution in the measured or modelled vorticity fields and the measurement of vorticity fluxes.

The present work is aimed at investigating the mechanisms associated to the vorticity fluxes in a flow where turbulence is generated by randomly placed cylinders. To achieve this goal two experimental tests were carried out: i) flow around an isolated cylinder (test I) and ii) flow within a random array of cylinders (test A). The experimental databases were acquired with a 2D Particle Image Velocimetry system (PIV) with spatial resolution that allows the computation of vorticity fields and vorticity fluxes.

Figure 1a shows the decay of the longitudinal vorticity flux across the line $x_1 = x^{(i)}$ between the cylinder axis, $x_2 = 0$, and $x_2 = d$ for the cylinder of test I and two cylinders in each measuring gap of test A. The decrease of the vorticity flux is more pronounced in the near wake of the cylinders within dense patches while in cylinders of sparse patches the longitudinal flux is similar to that of the isolated cylinder.

Figure 1b presents the lateral flux of vorticity across the plane $x_2 = 0$ (cylinder axis) as function of the normalized distance downstream of the cylinder base, x/d. The shape of $F_{2\omega_3}(x_1, x_2 = 0)$ is similar for all the cylinders.

One concludes that the pronounced reduction of the longitudinal vorticity flux in the wakes of cylinders in

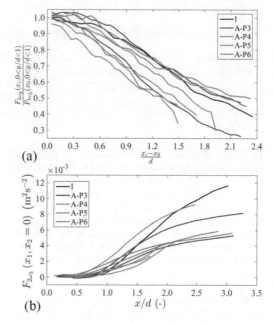

Figure 1. a) Ratio of longitudinal vorticity flux at $x^{(i)}$ and the initial point, $x^{(0)}$. b) Lateral vorticity flux $F_{2\omega_3}(x, y = 0)$.

dense patches is not accompanied by an increase of the lateral flux of vorticity across the symmetry axis of the cylinders, confirming the existence of vorticity cancellation mechanisms. The vorticity cancellation within random arrays of cylinders might be caused by the mechanisms associated to straining, by turbulent transport and by the transformation of the work of the pressure-strain correlation. Further research is needed to distinguish the role of each process in the vorticity cancellation within random arrays of cylinders.

ACKNOWLEDGMENTS

This work was partially supported by FEDER, program COMPETE, and by national funds through Portuguese Foundation for Science and Technology (FCT) project RECI/ECM-HID/0371/2012.

Effect of downstream channel slope on numerical modelling of dam break induced flows

A.E. Dinçer, Z. Bozkuş & A.N. Şahin
Middle East Technical University, Ankara, Turkey

ABSTRACT

In the present study, the numerical simulations of dam break induced flows are performed by using various models. The numerical models used in the study are laminar, large eddy simulation (LES) and Reynolds-Averaged Navier-Stokes (RANS) equations with k-ε turbulence model. In addition a recently developed Smoothed Particle Hydrodynamics (SPH) code is also used to simulate dam break problem. For the validation of the numerical approaches, a recently published experimental study of LaRocque et al. (2013) is used. In the experimental study, an idealized dam break problem in a sloped channel is investigated. They used a channel having 7.31 m length, 0.18 m width and 0.42 m depth. The bottom slope of the channel was 0.93%. 3.37 m from the upstream end, there was a wooden gate. This wooden gate was lifted suddenly in order to simulate instantaneous dam failure. For the velocity measurements, probes were placed in a horizontal position, 0.045 m above the channel bed. In this study, only the simulations of most upstream and most downstream measurement points from the wooden gate are presented. In addition, the simulation results for water surfaces are not presented. For the results of these simulations one can see the other studies of the authors about this subject.

The main scope of this study is to investigate the effect of downstream channel. First, the channel with a 0.93% slope, used in the experimental setup of LaRocque et al. (2013), was simulated. Then, the channel was assumed as horizontal and the simulations were repeated. Although the change in slope is necessary to investigate the effect of downstream channel, only a slope of 0.93% could have been used in the present study due to the lack of available experimental data.

The dimensionless parameter used in the study are: $X' = x/h_0$, $X'' = x/(t(gh_0)^{0.5})$, $V' = v/(gh_0)^{0.5}$ and $T = t(g/d)^{0.5}$. The numerical simulations compared with the experimental data for $X' = 2.00$ and $X' = 4.00$ which are the most upstream and most downstream points. The simulations of laminar flow, SPH, RANS and LES models are done for an initial reservoir head of 0.30 m. In the mesh-based simulations, the wall boundary with no-slip condition was used for the upstream end of the reservoir and for the bottom of the channel. In addition, symmetry boundary condition was used for the top of the computational area, and outflow boundary condition was used for the downstream end of the reservoir. In the SPH simulations, the initial distance between water particles is taken as 0.01 m. 10110 water particles and 1861 boundary particles are used in the simulations. The smoothing length is taken as 0.01 m. Both mesh-based and SPH computations are stopped after 2.5 s.

The simulation results show that when there is even a very small channel slope, the results of k-ε turbulence model deviate from the other simulation results and experimental data. Here it should be noted that this deviation is still fewer than 10% which is an acceptable limit for a numerical simulation. However, when there is no slope, the results of k-ε turbulence model are close other numerical models. In fact, there are no suggestions that propose the channel slope may affect the correctness of k-ε turbulence model in literature. Therefore, the authors avoid giving such a conclusion, but continue studying this phenomenon further.

REFERENCE

LaRocque, L., Imran, J., and Chaudhry, M. (2013). "Experimental and numerical investigations of two-dimensional dam-break flows." J. Hydraul. Eng.

Analysis of dotation discharge impact at a fishway entrance via numerical 3D CFD simulation

J. Klein & M. Oertel
Hydraulic Engineering Section, Civil Engineering Department, Lübeck University of Applied Sciences, Lübeck, Germany

ABSTRACT

Due to the European Water Framework Directive WFD (2000) all anthropogenic modified water bodies must be retreated into good ecological conditions. Thereby, main criteria to identify river's patency are functional fish step systems to allow up-and downstream fish migration. Several fish step systems can be used to produce hydraulic flow systems with fish climb capability. Technical structures (e.g. vertical slot pass) can be arranged on small footprint areas next to weirs with included hydro-power systems. Nature-like solutions shall provide a natural flow situation and an attractive structure within the environment. These can be e.g. crossbar block ramps or ramps with interlocked blocks.

The success of a fishway depends on two criteria: (1) fish passage capability, and (2) fishway detect ability. Only if both criteria are fulfilled a fishway can be considered as successful. To design a passable fishway structure defined geometric value limits due to hydraulics and fish physiology must be considered. Additionally, the entrance of the fishway must be passable as well as detectable. The present paper deals with a planned fishway in northern Germany. Due to tidal influenced flow parameters, the fishway outlet was designed to be orthogonal to the downstream river system. Hence, negative influences concerning fish entering the structure as well as ship movement can be assumed. Besides passable flow velocities and a minimum of turbulence at the entrance area, the appearance of a continuous leading flow with velocities larger than 0.2 m/s are important parameters to guarantee a successful structure. These parameters were taken into account to design a fish friendly entrance. Furthermore, public authorities gave limitations of maximum mean inlet velocities because the downstream water body is a small federal waterway with ship traffic where obstruction has to be avoided. The entrance of the fishway as well as 100 m of the downstream water body is analysed via numerical 3D CFD simulations. The main aim is to verify the hydraulic flow situation at the inlet zone of the fishway. Investigations include three different inlet structures and results lead to an optimized design of the fishways entrance.

Concluding, numerical simulation provides a sufficient tool to evaluate hydraulic issues during the planning phase. The present investigation shows a possibility to improve the flow situation at the inlet of a fishway which cannot be a design guideline solution. However, attention should be paid to quality components which have a main influence on success and cost-benefit of numerical simulations. But hydraulic investigations can never evaluate fish's behaviour. Therefore it is advisable to investigate the fish's behaviour in the investigation area carefully to evaluate the structure's final design. Experiences showed that subsequent adjustments at fishways are common. Thus, it is also advisable to construct a flexible inlet structure to react on fishes' behaviour.

Characterization of flow resistance in a floodplain for varying building density

S. Guillén-Ludeña, D. Lopez, E. Mignot & N. Rivière
Université de Lyon, INSA de Lyon, Laboratoire de Mécanique des Fluides et d'Acoustique, France

ABSTRACT

In extreme flood events land occupation in floodplains becomes a major issue as it highly influences the flow resistance and therefore, the hydrodynamic processes. The present work aims at identifying, as the land occupation increases i) the evolution of the flood risk associated with a high return period flood event and ii) the relative magnitude of the resistance forces acting on the flow by both the bed friction and the drag induced by buildings. For these purposes, this paper presents an experimental methodology to assess the influence of both the bed-friction forces and the drag forces on the overall flow resistance. Also, the transition from a flow governed mostly by bed-friction forces to a flow in which the drag forces due to the obstacles are predominant is studied herein.

The strategy consists in measuring the drag force (F_d) that the flow exerts on one obstacle located at the middle of a cell (see Fig. 1). Once the value of F_d is known, the bed-friction force (F_b) can be obtained by applying momentum balance to the measured cell, which performed along the streamwise direction and under uniform flow conditions, results in the following expression:

$$F_d + F_b + W = 0 \qquad (1)$$

where F_d is the drag force, F_b is the bed-friction force, and W stands for the weight component of the water.

The purpose is to obtain the evolution of the ratio F_d/F_b for increasing values of the spatial density of the obstacles (λ) and of the ratio d/L, where d is the flow depth and L is the distance between the centers of two consecutive obstacles (see Fig. 1). Also, the interaction between bed-friction and drag forces is studied by analyzing on the one hand, the influence of the

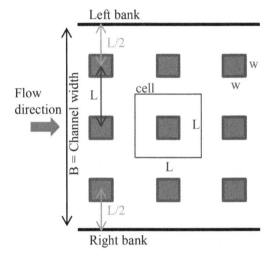

Figure 1. Schematic plan view of the obstacle distribution.

bed friction on the drag coefficient and on the other hand, the influence of the obstacles on the bed-friction coefficient.

The presented methodology characterizes the transition from a flow whose resistance is governed by the bottom friction (for a limited built land occupation), towards a flow whose resistance is governed by the drag forces on buildings (for a high built land occupation).

ACKNOWLEDGEMENTS

This project is founded by the ANR project FLOWRES (ANR-14-CE03-0010)

Turbulent dispersion in bounded horizontal jets. RANS capabilities and physical modeling comparison

D. Valero
FH Aachen University of Applied Sciences, Aachen, Germany
University of Liege (ULg), Liège, Belgium

D. B. Bung
FH Aachen University of Applied Sciences, Aachen, Germany

M. Oertel
FH Lübeck University of Applied Sciences, Lübeck, Germany

ABSTRACT

Accurate prediction of contaminant concentrations can yield improved outfall structures design, reducing economical expenses and environmental impact.

Conventionally, cost intensive physical modeling or simplified integral approach models have been employed despite their drawbacks and limitations.

When using a numerical model, practitioners are usually attracted to the integral method approach, which is subject to considerable restrictions due to model's hypothesis. However, they hold an extensive experience and validation.

With increasing computer capacities, more complex models have arisen. 20 years ago it was not possible to simulate complex river reaches – nowadays, new hardware advances make it feasible at affordable costs. Thus, Reynolds Averaged Navier-Stokes (RANS) equations appear to be the more common approach in Computational Fluid Dynamics (CFD) when a 3D approach is desired. Large Eddy Simulations (LES) can be already performed and may yield to more accurate and detailed results. However, modeling large domains of interest such as in environmental discharges problems or complex multi-physics might be restricted to RANS modeling. Additionally, the advanced level of competence required to run a LES is an obstacle to its widespread application.

In the present paper, two jet setups have been studied by means of 3D Reynolds Averaged Navier-Stokes equations and experimental modeling. Both jet cases correspond to turbulent horizontal jets, bounded by the channel bed, which might be found in common environmental discharges. Three of the most widely employed turbulence models up to date have been investigated (namely standard k-ε, RNG k-ε and k-ω), analyzing their performance on the jet trajectory estimation. For the best performance's model, RNG k-ε and for both jets cases of the present study, analysis has been extended to the turbulence diffusion estimation by defining a turbulent Schmidt number.

It has been shown that RNG k-ε is able to better reproduce jet trajectories despite some differences while definition of a turbulent Schmidt number becomes necessary in order to account for turbulence dispersion. Usage of a turbulent Schmidt number - when properly selected – reduces relative errors in the jet boundary width one order of magnitude, generally yielding to errors less than 10 % for such variable.

It is remarkable that turbulence model selection induces a major change in overall result and thus special care is necessary. Additionally, when using k-ω, which has shown larger unsteadiness, different turbulent Schmidt numbers might be expected. This unsteadiness can surpass the effect of the real turbulent dispersion precluding use of any turbulent Schmidt value.

Consequently, use of different turbulent Schmidt numbers may help assess uncertainty of RANS numerical modeling in the design of complex outfall structures. Capabilities of the employed models are wider than integral model approach with a reasonable computational cost (when compared to more cost intensive techniques, i.e.: LES).

Modelling methane emissions from Vilar reservoir (Portugal)

R.F. Lino, P.A. Diogo, A.C. Rodrigues & P.S. Coelho
MARE, FCT-UNL, Portugal

ABSTRACT

Hydroelectricity has been highly implanted worldwide due to its high energy production capacity but although it is a renewable energy source, the construction of hydropower dams is far from being a "clean" solution: landscape destruction, flooding and altering the natural course of rivers caused by dams, as well as water quality problems are some of the issues related to dam construction. Furthermore, the decomposition of the sediments in the benthic zone of reservoirs may induce the production of greenhouse gases which end up being released to the atmosphere.

This study focused on the quantification of methane emissions to the atmosphere from Vilar reservoir by using the model CE-Qual-W2. The importance of methane emissions is strongly associated with the fact that it has a global warming potential 25 times higher than carbon dioxide, thus providing more negative effects in smaller quantities than CO_2.

Early results indicate no emissions and practically no methane production at all. According to Beaulieu et al. (2014), methane emissions from reservoirs in tempered climates are usually below 50 mg C.m-2d-1 and in tropical climates between 50 and 150 mg C.m-2d-1, suggesting that modelling results obtained for the Vilar reservoir may be an overestimation of methane releases pointing out the need for further analysis.

Considering the reliability and experience in the use of the CE-Qual-W2 model, it is considered that although field data limitations are yet a problem to be tackled, the implementation of this model is surely a way to go towards a better understanding of the potential contribution from reservoirs to GHG emissions.

REFERENCE

Beaulieu, J. J., Smolenski, R. L., Nietch, C. T., Townsend-Small, A. and Elovitz, M. S., 2014. High Methane Emissions from a Midlatitude Reservoir Draining an Agricultural Watershed. *Environmental Science & Technology* 48, pp. 11100–11108.

The impact of stormwater overflows on stream water quality

T. Julínek & J. Říha
Institute of Water Structures, Brno University of Technology, Czech Republic

ABSTRACT

In the past, municipalities were traditionally provided with combined sewer systems. Over the decades, newly built urban areas have been appended to the existing sewer mains, causing their frequent overloading. Various technical and environmental measures have been adopted in order to provide sustainable stormwater management (Malmqvist 2000). Within the comprehensive reconstruction of the sewer network in the city of Brno, the improvement of existing stormwater separators (overflows) and also the design of new ones are planned. The rehabilitation of sewerage also involves the design of stormwater retention tanks, which lower the peak discharges in the sewers and so decrease the released amount of polluted water to receiving rivers via stormwater separators.

The purpose of the presented study was to assess the changes over time in the concentration of six water quality indicators (BOD, COD, N-NH$_4$, N$_{tot}$, P$_{tot}$ and suspended solids SS) in the principal rivers in the area of the city of Brno. These rivers are the Svratka and Svitava rivers and the Leskava and Ponavka streams, along which principal sewer mains are located and equipped with combined sewer over-flows. The results of the study describe the development of maximum concentrations of the above-mentioned indicators along the streams and also the total average annual mass load released into surface streams. Numerical modelling of the stream water quality was carried out using MIKE11 computer code. The main objective of the study was to assess the effect of measures applied to the sewer system on the water quality in streams. Two scenarios were resolved, namely the present state and a proposed set of improvements.

The results of modelling showed that:

- at present, the stream water quality during extreme rainfall events in the Brno area exceeds the limits prescribed by immission water quality standards (Decree23 2010) by more than 10 times,
- the improvements will considerably decrease the amount of pollution released. At the most critical reaches along the Svratka and Svitava rivers the concentration of pollutants will be reduced by approx. 50 % (Tab. 1),
- in the case of a simulated extreme event the water quality standards in the analyzed rivers will be violated for a period of more than two hours.

The model results show that the measures proposed for the sewer system (retention tanks, overflows, etc.) will reduce the peak concentrations of individual monitored indicators of water quality by 45 to 90 % of those recorded for the present state. The occurrence time of the concentration peaks varies from 10 minutes in the upper river reaches up to several hours at the reaches close to the junction of the Svitava and Svratka rivers.

Table 1. Maximum calculated concentrations of pollutants in [mg/l] and multiple of immission limits (IL) – present state at discharge Qa.
Maximum concentration in [mg/l] / multiple of immission limit.

Indic.	Svratka*	Svratka**	Svitava	Leskava	IL
BOD$_5$	116/19	65/11	62/10	4.0/0.7	6
COD	293/8.4	193/5.5	195/5.6	22.0/0.6	35
N-NH$_4$	6.0/12.0	5.6/11.1	11.1/22.3	0.3/0.6	0.5
N$_t$	26/3.2	16/2.0	22/2.8	5.5/0.7	8
P$_t$	4.8/31.9	2.5/16.4	2.1/14.1	0.4/2.7	0.15
SS	662/27	351/16.4	281/14.1	10/0.4	25

Note: Svratka* at km 44.11, Svratka** at km 40.19, the River Svitava at km 11.2 and the Leskava River at km 1.15. IL – immission limit.

REFERENCES

Decree23. 2010. Decree No. 23/2010 Coll. on indicators and values of permissible pollution of surface water and wastewater, details of the permit to discharge wastewater into surface water. (Czech national legislation).

Malmqvist, P.A. 2000. Sustainable storm water management – some Swedish experiences. Journal of Environmental Science and Health Part A A35(8): 1251–1266.

Groundwater quality of Feriana-Skhirat in central Tunisia and its sustainability for agriculture and drinking purposes

I. Hassen
*Modeling Hydraulic and Environmental Laboratory, National Engineering School of Tunis,
University of Tunis El Manar, Tunisia*

F. Hamzaoui-Azaza
*Research Unit of Geochemistry and Environmental Geology, Faculty of Science of Tunis,
University of Tunis El Manar, Tunisia*

R. Bouhlila
*Modeling Hydraulic and Environmental Laboratory, National Engineering School of Tunis,
University of Tunis El Manar, Tunisia*

ABSTRACT

Located in North-Western of Kasserine (Central Tunisia), the deep aquifer of Feriana-Skhirat is among the most important groundwater in Kasserine aquifer system. It is surrounded by mountains of Jb Goubel, Jb Serraguia and Jb Feriana.

To carry out the study, a total of 13 samples were collected during Januray-February 2014 from different location in the groundwater of Feriana-Skhirat. Sample collection, handling and storage followed standard procedures recommended to ensure data quality and consistency. A set of 12 physicochemical parameters and majors ions: (Temperature, pH, salinity, electrical conductivity, Na, K, Mg, Ca, HCO_3, Cl, SO_4 and NO_3) were conducted in the laboratory of the Centre of Hydrogeology and Geothermic (CHYN) of the University of Neuchatel (Switzerland). All major and minor ions are within the acceptable limit of WHO for drinking water (World Health Organization, 2008).

The understanding of the geochemical behavior was achieved by using conventional classification techniques and correlations. Interpretation of analytical data showed that the water facies gradually change from $SO_4^{2-}-Ca^{2+}-Mg^{2+}$ to $SO_4^{2-}-Na^+$ type and are controlled by rock-water interaction. The first water type results mainly from carbonate minerals precipitation and dissolution of gypsum minerals which are relatively abundant within the Upper Miocene sequences. While, the $SO_4^{2-}-Na^+$ type water, represented by FS4 and FS11, is controlled by rock water interaction. This facies is highly mineralized comparing to the other wells and results from the substitute of calcium by sodium through cation exchange reaction may be the main process controlling the geochemistry of Feriana-Skhirat aquifer. Such result was also identified by several ions ratios and correlation parameters. Furthermore, in the present study, the water samples of Feriana-Skhirat are saturated/under saturated with respect to Dolomite and Calcite ($-1.7 < SI_{(Dolomite)} < 1.3$ and $-0.8 < SI_{(Calcite)} < 0.7$). With such results, carbonate precipitation may occur systematically for the majority of groundwater samples. Consequently, the precipitation of these minerals causes equilibrium in Ca^{2+} concentrations and leads to the dissolution of evaporate materials (Gypsum and anhydrite). Otherwise, there is an under saturation state with Gypsum, Halite and Anhydrite ($-1.5 < SI_{(Gypsum)} < -0.73$ and $-7.4 < SI_{(Halite)} < -6.08$ and $-1.75 < SI_{(Anhydrite)} < -0.9$) indicating dissolution of these evaporate minerals. Therefore, the soluble component for these evaporate minerals Na^+, Cl^-, Ca^{2+} and SO_4^{2-} concentrations are not limited by mineral equilibrium.

As for groundwater assessment, irrigation indices (SAR, %Na) indicate that most of the groundwater samples of Feriana-Skhirat aquifer were suitable for agriculture purpose. The Wilcox diagram shows that Feriana-Skhirat groundwater samples are excellent to good for irrigation except two samples that fall in the field of doubtful to unsuitable. Otherwise, Richards diagram indicates that Feriana-Skhirat groundwater can irrigate different types of soil with little hazard of exchangeable sodium with an exception of two samples that indicates very high salinity and medium alkalinity hazard. Thus, this water can be used for plants that tolerate high salt concentration. The calculation of the WQI of this aquifer indicates that 100% of groundwater samples of Feriana-Skhirat aquifer defined "Excellent Water".

REFERENCE

Hamzaoui-Azaza F, Ameur M, Bouhlila R, Gueddari M. 2012. Geohemical Characterization of groundwater in a Miocene Aquifer, Southeastern Tunisia. Environ Eng Geocsci 18:159–174.

Modelling the transport and decay of microbial tracers in a macro-tidal estuary

A.A. Bakar, R. Ahmadian & R.A. Falconer
School of Engineering, Cardiff University, Cardiff, UK

ABSTRACT

The Loughor Estuary is a macro-tidal coastal basin, which is located along the Bristol Channel, in the South West of the U.K. This estuarine region can experience severe coastal flooding during high spring tides, as a result of the maximum spring tidal range of 7.5 m, occurring near Burry Port. As the Loughor Estuary and the surrounding coastal waters provide natural sites for recreational bathing and shellfish beds, with water quality in this water basin being important for compliance with the designated standards, as to secure the safety of the bathing and shellfish harvesting industries. The estuarine system, however, potentially gets impaired by receiving bacterial overloads from catchments through combined sewer overflows and rivers (Kay et al. 2008), besides the extreme inter-tidal flooding at Llanrhidian saltmarsh, and from the lower industrial and residential areas at Llanelli and Gowerton.

A two-dimensional hydro-environmental model has been developed to simulate the hydrodynamic and turbulence processes in the Severn Estuary and Bristol Channel. The model has been refined and extended to include the highest water level, covering the inter-tidal floodplains of the estuary, and with integration of LiDAR data with the bathymetry of the model. Four different microbial tracers – *Serratia marcescens*, *Enterobacter cloacae*, *MS2*, and *φX174* phages, were isolated from the seawater and sewage and could be inactivated by extremes of pH, heat and solar radiation (Wyer et al. 2014), are released at the Loughor Estuary from four different locations.

The transport of these tracers were calibrated for advection and diffusion, by considering the bed features of the inter-tidal floodplains and dispersion due to turbulence and flushing during slack tides by river discharges. The decay of the released microbial tracers were modelled as the simple first-order rate functions, with the time required to exponentially inactivate 90% of the mass of the microbial tracers being represented

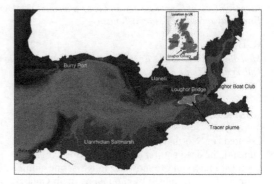

Figure 1. Released microbial tracer in the Loughor Estuary.

by the T_{90} value (Schnauder et al. 2007). The T_{90} value was calibrated for different ranges between day and night, with the higher decay rate during day times being used to represent the effects of solar radiation. The results from the transport and decay model have been compared with measured data at five different sampling locations. The refined model developed as a part of this study was used to improve on the understanding of the water quality processes and the potential sources of bacterial pollution in the estuary.

REFERENCES

Kay, D., Kershaw, S., Lee, R., Wyer, M.D., Watkins, J. and Francis, C., 2008. Results of field investigations into the impact of intermittent sewage discharges on the microbiological quality of wild mussels (Mytilus edulis) in a tidal estuary. Water Research, 42, 3033–3046.

Schnauder, I., Lin, B. and Bockelmann-Evans, B.N., 2007. Modelling faecal bacteria pathways in receiving waters. Proceedings of the ICE – Maritime Engineering, 160, 143–153.

Wyer, M., Kay, D., Naylor, S., Thomas, S. and Fracchiolla, A., 2014. Loughor Estuary Microbial Tracer Study Logistics: Field Study Outline.

Building water and chemicals budgets over a complex hydrographic network

V. Carbonnel
Department of Water Pollution Control, CP208, Université Libre de Bruxelles (ULB), Belgium

N. Brion, M. Elskens & P. Claeys
Laboratory of Analytical, Environmental and Geo-chemistry, Vrije Universiteit Brussel, Belgium

M.A. Verbanck
Department of Water Pollution Control, CP208, Université Libre de Bruxelles (ULB), Belgium

ABSTRACT

The Brussels Metropolitan Community (BMC) is a densely populated, trans-regional area of circa 800 km² (about 2,000,000 inhabitants), over which Brussels economy and urbanization will thrive and expand in the coming decades. The interconnected system composed of the Zenne and the Charleroi-Brussels-Scheldt ship canal (Fig. 1), which supports the complex hydrographic network in the BMC, constitutes the backbone for the sustainable development of the economy in this area.

Profoundly modified over the last two centuries to address several issues such as flooding and water pollution, this hydrographic network constitutes a paradigm of a complex system submitted to multiple types of human perturbations impacting the hydrologic functioning, the water courses, the water budget, and also, directly or indirectly, the water quality. In addition, as it spans over the three Belgian Regions

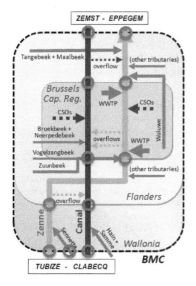

Figure 1. Scheme of the water transit lines in the BMC. Red circles represent the limits of the box model (see text).

Table 1. List of the followed parameters and tracers.

Group	Parameter	Source
Basic	Conductivity, pH, O_2, suspended matter	
Tracers	K, B	Sewage
Nitrogen	NH_4	Sewage/in stream
	NO_3	Agriculture/in stream
Metals	Pb, Zn, Cd, Cu, Ni, Hg	Industry/urban runoff
Isotopic	$\delta^{11}B$	Sewage
	$\delta^{15}N\text{-}NH_4$	Sewage/in stream
	$\delta^{15}N\text{-}NO_3$, $\delta^{18}O\text{-}NO_3$	Agriculture/in stream
	$\delta^{206}Pb$, $\delta^{207}Pb$, $\delta^{208}Pb$	Industry/urban runoff
	$\delta^{66}Zn$	Industry/urban runoff
Pesticides	Diuron, isoproturon	Agriculture
PAH	Fluoranthene, benzo(k)fluoranthene, benzo(b)fluoranthene	Urban runoff, Urban runoff, Urban runoff

(Flanders, Brussels and Wallonia, Fig. 1), the management of this system is divided in sections which are under the supervision of different organisms and administrations.

Our general objective was to study the propagation of the anthropogenic disturbances in this aquatic system that goes through such a densely populated and economically active area, with multiple connections and sources of pollutants. We thus established a box-model representation of water and polluting chemicals budgets for the interconnected Zenne and Canal on the whole BMC area, with the inner nodes set at the regional boundaries (red circles in Fig. 1).

As identifying and quantifying every individual chemical is a tremendous and costly task, this aim was achieved by combining hydrological and sediment dynamics analyses with the use of selected traditional and innovative chemical tracers representing different types of human activity: agriculture, industries and urban surface runoff and sewage (Table 1).

Data were retrieved from: 1) several large existing datasets compiled in each region, 2) hydrological and suspended particle transport modeling, and 3) field campaigns conducted during different seasons at box model boundaries and, for background conditions, at the sources of water streams representative for the catchment.

Evaluating the ecological restoration of a Mediterranean reservoir

P. Sidiropoulos & M. Chamoglou
Management Body of Ecodevelopment Area of Karla - Mavrovouni – Kefalovriso – Velestino, Greece

I. Kagkalou
School of Engineering, Department of Civil Engineering, Democritus University of Thrace, Sector of Hydraulics, Xanthi, Greece

ABSTRACT

The restoration of Lake's Karla environment and ecosystem is studied through the evaluation of its current status. Lake Karla was one of the most important wetlands of Greece with many benefits not only to biodiversity conservation, to water balance of the watershed, but also to local economy from fisheries. Its drainage, in 1962, created a series of environmental problems and led to the local economy contraction. The Lake Karla reconstruction project has begun since 2000 and has not been implemented yet, although it was planned to be completed in 2012. The construction of a reservoir at the lowest part of Lake Karla watershed will be among the parts of this project, which will be supplied from the near by Pinios River and the runoff of the surrounding mountains. This newly re-established water body is considered a vital aquatic ecosystem as it is listed in the network of Natura 2000 and has been characterized as a Permanent Wildlife Refuge by Greek Law.

On the other hand in semi-arid and arid areas, including this site located in the Mediterranean region, reservoirs often represent quite extreme environments because of pronounced changes in water levels with associated periods of very high nutrient concentrations while sediment enriches water column with nutrients.

Regarding at the project design, technical studies, recommended draining the lake via the Karla Tunnel and building a smaller reservoir instead of the natural lake for flood protection and for the revelation of agricultural fields. The re-constructed Lake Karla occupies the lowest part of the former Lake Karla. It lies between latitude 39°26'49" to 39°32'03' N and longitude 22°46'47" to 23°51'50" E and has a surface of 38 km^2. It is characterized by its shallow depth with a maximum water depth of 4.5 m and a mean depth of 2 m.

A monitoring program was deployed focusing on the water quality along with the trophic conditions, the hydrological budget and the new wetland's biodiversity. The European Directives (i.e the Water Framework Directive and the Habitat Directive) were used as guidelines for the applied methodology, for the characterization of the water body and for the assessment of its status.

The new lake Karla is exposed to point and diffuse pollution sources leading already to a progressive eutrophication. Lake Karla receives all types of pollution acting as a sink for pollutants, suspended matter and toxics. The new reservoir is characterized as hypereutrophic with the frequent occurrence of algal blooms dominated by cyanobacteria. Noteworthy is that the bird fauna has been increased and the area is still of paramount importance for migratory, wintering and breeding waterfowl, waders and birds of prey while fish population is dominated by cyprinids as it happens in shallow warm lakes and reservoirs.

The results indicate that the delay of implementation of Lake Karla reconstruction project, the violation of Environmental Terms and the lack of environmental policy are the most important cause of degradation. The human induced pressures are still so intense on Lake Karla and its ecoregion thus it is very probable to generate irreversible conditions.

REFERENCES

European Commission, 1992. Directive 92/43/EEC, Conservation of natural habitats and of wild fauna and flora. *Official Journal of the European Communities* No L 206/7/1992.

European Commission, 2000. Directive 2000/60/EC of the European Parliament and the Council of the 23 October 2000 establishing a framework for Community action in the field of water policy. *Official Journal of the European Communities*. L 327:1–73.

Spatio-temporal interaction of morphological variability with hydrodynamic parameters in the scope of integrative measures at the river Danube

M. Glas, M. Tritthart, M. Liedermann, P. Gmeiner & H. Habersack
*Christian Doppler Laboratory for Advanced Methods in River Monitoring, Modelling and Engineering;
Institute of Water Management, Hydrology and Hydraulic Engineering; Department of Water,
Atmosphere and Environment; University of Natural Resources and Life Sciences, Vienna, Austria*

ABSTRACT

Hydrodynamic parameters, determined by 3D numerical models, provide a basis for further simulation tools, such as sediment transport and habitat models. By implementing morphodynamic models which are covering temporal scales of several decades, a restriction to relatively simple numerical codes is inherited due to their more feasible computational times. On the other hand, the use of sophisticated 3D numerical codes (e.g. Tritthart, 2005) restricts simulation times to ranges far below those time scales (Tritthart et al., 2011). When employing sophisticated codes, measures in rivers are often assessed by numerical models only right before and after the measure, due to the high demand in data, skills and time. In this study, a sensitivity analysis was presented, capturing the interaction of hydrodynamics, sediment transport and morhodynamics. A three kilometer long study site, located at the Danube East of Vienna, was at the focus of this study, due to the implementation of large scale integrative measures (alternative groyne layout, granulometric bed improvement, river bank restoration and side arm reconnection), combined with a detailed and long term monitoring program. As a first result, flood affected morphologies were identified to be characterized by a different distribution of hydrodynamic parameters in comparison to morphologies which were not affected by extreme events. Furthermore, a general effect of the integrated measures was prevalent, visible in morphology and hydrodynamic parameters (water depth, depth averaged flow velocity and bed shear stress). Finally, variability in hydrodynamic parameters within different pre-measure models was found, due to the variation in morphology. Figure 1 exemplarily presents the variability of bed shear stress along the river axis. Maximum ranges of modelled hydrodynamic parameters were 0.81 m, 0.17 ms^{-1} and 3.05 Nm^{-2} for modelled water depth, depth averaged flow velocity and bed shear stress, respectively.

The authors propose to investigate this variability in a scenario-based sensitivity study by implementing at least two scenarios with low and high bed levels. Concerning bed shear stress, the implemented integrative measures within this study led to a reduction of up to 3 Nm^{-2} for a scenario with relatively low bed levels and a reduction of 1 to 5 Nm^{-2} for a scenario with relatively high bed levels. This difference yields a potential effect on sediment transport models – and other simulation tools like habitat models – and therefore should be considered for planning and assessing future integrative measures with the help of complex numerical codes (e.g. 3D numerical models).

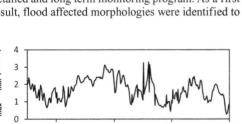

Figure 1. Variability of modelled bed shear stress τ over time (2007/03 to 2012/09) along a longitudinal section (river axis).

REFERENCES

Tritthart, M. 2005. *Three-Dimensional Numerical Modelling of Turbulent River Flow using Polyhedral Finite Volumes*. Vienna: Wiener Mitteilungen Wasser-Abwasser-Gewässer.

Tritthart, M., Liedermann, M., Schober, B. & Habersack, H. 2011. Non-uniformity and layering in sediment transport modelling 2: river application. *Journal of Hydraulic Research* 49(3): 335–344.

Analysis of contributions and uncertainties of fish population models for the development of river continuity concepts in the river basin Ruhr, Germany

D. Teschlade & A. Niemann
Institute of Hydraulic Engineering and Water Resources Management, University of Duisburg-Essen, Essen, Germany

ABSTRACT

Rivers belong to the most stressed and intensively human-influenced ecosystems in the world. Numerous anthropogenic obstructions like water diversions for using hydropower and other industrial purposes, irrigation and domestic uses are the reason why most of the world's largest rivers are nowadays fragmented by dams (Jager et al., 2001).

In order to achieve the goals of the Water Framework Directive (the "*good ecological status*"), numerous river restoration measures have been implemented in recent years. In consideration that not all dams may be removed there is an urgent need to develop a procedure to prioritize most critical barriers. In this case one of the main current challenges in restoration fish migration is the use of large databases of biological monitoring surveys to help environmental managers prioritizing restoration measures to address the connectivity problem more strategically.

This project provides a modeling framework to develop the methodological basis for the analysis of ecological continuity needs by using the method of Radinger and Wolter (2015). They combined species habitat models (MaxEnt; (Elith et al., 2011)) with species dispersal models (FIDIMO; (Radinger et al., 2014)) to show the effects of habitat suitability, dispersal, and fragmentation on the distribution of river fishes (Radinger and Wolter, 2015).

During the first research about the feasibility of this project and first modelling results it becomes clear, that population models are in general suitable for the development and derivation of prioritized river continuity concepts. The developed target fish species model for *thymallus thymallus* (grayling) provided "excellent" (AUC > 0.80) predictions. Furthermore the MaxEnt simulation show that the single most influential predictor for the presence/absence of the investigated species was the riverbed width (54.9%). A threshold effect was evident at approximately eight meter riverbed width for *thymallus thymallus* (grayling). In addition to that also special structures in form of wood, latitudinal variance and flow diversity are predictors for the species.

After creation of habitat suitability maps for each fish species we will start to analyze fish dispersal. Combining both habitat suitability and species dispersal maps we will be able to evaluate barriers in terms of the achievement of habitats.

Within the project the fish dispersal model FIDIMO will be optimized in the field of ecological continuity. Particularly a more precise description of passability of technical fishpasses should be implemented in the model. For this, the existing investments are analyzed and evaluated. To improve the method we will start to evaluate various migration obstacles systematically to see their cumulative effect on fish populations. A special focus is placed on technical fish passes and downstream fish protection devices at hydropower plants to analyze the dynamics of fish populations. One of the main goals is to use existing data only to ensure transferability to different counties or river basins authorities and to aid the implementation of the WFD. For this purpose, various scenarios of river continuity will be simulated.

REFERENCES

Elith, J., Phillips, S. J., Hastie, T., Dudík, M., Chee, Y. E. & Yates, C. J. 2011. A statistical explanation of MaxEnt for ecologists. *Diversity and Distributions*, 17, 43–57.

Jager, H., Chandler, J., Lepla, K. & Van Winkle, W. 2001. A Theoretical Study of River Fragmentation by Dams and its Effects on White Sturgeon Populations. *Environmental Biology of Fishes*, 60, 347–361.

Radinger, J., Kail, J. & Wolter, C. 2014. Fidimo – A free and open source GIS based dispersal model for riverine fish. *Ecological Informatics*, 24, 238–247.

Radinger, J. & Wolter, C. 2015. Disentangling the effects of habitat suitability, dispersal, and fragmentation on the distribution of river fishes. *Ecological Applications*, 25, 914–927.

Hydrologic and hydraulic design to reduce diffuse pollution from drained peatlands

S. Mohammadighavam & B. Kløve
*Water Resources and Environmental Engineering Research Group,
Faculty of Technology, University of Oulu, Finland*

ABSTRACT

Peatland ditching for forestry, agriculture and peat extraction in boreal regions has a significant effect on water quality and quantity (Fig. 1) which needs sustainable and careful management (Dunn & Mackay 1996, Kløve 2001, Marttila & Kløve 2008). Several methods such as wetlands, peak runoff control and chemical water treatment have been introduced to limit the environmental impacts .(Tammela et al. 2010), but the overall design still lack basic hydrologic/hydraulic information for optimal and cost-efficient design. Therefore, the objective of the presented research is to develop new methods based on more refined hydrologic information and hydraulic design that will better predict, control and improve water management solution for ditched peatlands to reduce leaching and diffuse pollution to watercourses.

The study builds on field and laboratory experiments as well as hydrologic/hydraulic modelling. Five sites (3 forestry drained and 2 peat extraction areas) in northern Finland were selected and extensively observed (years 2011–2014). Peat physical properties and monitoring data on rainfall, runoff and groundwater levels were used to build hydrological model using DRAINMOD 6.1 to predict WTD fluctuations and drain runoff for short and long-term periods. Also several three-dimensional (3D) computational fluid dynamic (CFD) turbulence models were built using COMSOL Multiphysics 5.1 commercial Software to modify and optimize different parts of sedimentation

Figure 2. Full-scale around-the-end gravity-driven hydraulic flocculator under construction as part of drained water treatment facilities on a peat extraction field in northern Finland.

pond structures located at the drain outlets and used in water treatment. The obtained results (Saarinen et al. 2013, Mohammadighavam et al. 2015) were used to modify existing facilities and build a full scale gravity driven hydraulic mixer with 4 m width and 65 m (Fig. 2) length which is under investigation to improve the accuracy of CFD modelling.

REFERENCES

Dunn, S. M., & Mackay, R. 1996. Modelling the hydrological impacts of open ditch drainage. *Journal of Hydrology, 179*(1–4): 37–66.

Kløve, B. 2001. Characteristics of nitrogen and phosphorus loads in peat mining wastewater. *Water Research, 35*(10): 2353–2362.

Marttila, H., & Kløve, B. 2009. Retention of sediment and nutrient loads with peak runoff control. *Journal of Irrigation and Drainage Engineering, 135*(2): 210–216.

Mohammadighavam, S., Heiderscheidt, E., Marttila, H., & Kløve, B. 2015. *Optimization of Gravity-Driven Hydraulic Flocculators to Treat Peat Extraction Runoff Water. J.Irrig.Drain.Eng.,* 04015045.8

Saarinen, T., Mohammadighavam, S., Marttila, H., & Kløve, B. 2013. Impact of peatland forestry on runoff water quality in areas with sulphide-bearing sediments; how to prevent acid surges. *Forest Ecology and Management, 293*(0): 17–28.

Tammela, S., Marttila, H., Dey, S., & Kløve, B. 2010. Effect and design of an underminer structure. *Journal of Hydraulic Research, 48*(2): 188–196.

Figure 1. Sediment and nutrient leaching from ditched peatland to downstream watercourses.

Flow modelling and investigation of flood scenarios on the Cavaillon River, Haiti

A. Joseph & N. Gonomy
GNR-FAMV, Université d'Etat d'Haïti, Port au Prince, Haiti

Y. Zech & S. Soares-Frazão
Institute of Mechanics, Materials and Civil Engineering, Université Catholique de Louvain, Louvain-la-Neuve, Belgium

ABSTRACT

The vulnerability of riverine populations confronted to rivers floods and the socioeconomic consequences of floods remain a major concern for the Haitian public decision-makers. However, the use of modelling tools for flood prediction and mitigation purposes is not yet well established in the country. This paper presents the first results of an ongoing study on the Cavaillon River in Southern Haiti, with the aim of constructing a methodology to design a simple, efficient and flexible hydraulic model for Cavaillon river, based on the existing knowledge and on field surveys, and using free modelling tools, considering the limited available resources in the country.

The studied reach has a length of 25 km, between the Dory weir and Grand Place (Figure 1). At the start of the study, only a 30 m × 30 m DTM was available, without any bathymetrical data. Following a field survey (Joseph et al., 2015), 96 cross-sections were measured along this reach.

A one-dimensional model of the river reach was constructed based on the collected field data, and calibrated using available discharge and water level measurements (Figure 2). In particular, the bed friction was determined based on grain-size distributions

Figure 1. Localization of the studied river reach.

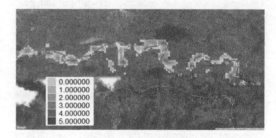

Figure 2. Water depth – $Q = 1000$ m^3/s.

obtained from the application of the Wolman Pebble Count Method (Wolman, 1954).

Floods scenarios were then run for discharges Q of 535 and 1000 m^3/s, respectively. For these two discharges, the water-depth map as illustrated in figure 3 and the extend flood map were generated. As expected the flood extent is considerably larger for the largest discharge and the water depth is significantly increased

Although there are some anomalies, the first simulations results and the flooded areas representation show the relevance of the methodology based on free software tools.

REFERENCES

Gonomy N. (2012). Projet d'études et élaboration de plans d'aménagement de trois bassins versants dans le cadre du programme de mitigation des désastres naturels en Haïti. Rapport d'études (in French).

Joseph A., Carlier d'Odeigne O., Zech Y., Gonomy N., Soares-Frazão S. (2015). Construction of one-dimensional model for the Cavaillon River, Haiti, from an in-situ survey of the bathymetry, e-proceedings of the 36th IAHR World Congress 28 June–3 July, 2015, The Hague, the Netherlands

Wolman G. (1954). A method of sampling coarse river-bed material. Transactions, American Geophysical Union, 35 (6), 951–956.

Simulating runoff generation and consumption for rehabilitation of downstream ecosystems in arid northwest China

C. He
Key Laboratory of West China's Environmental System (Ministry of Education), College of Earth and Environmental Sciences, Lanzhou University, Lanzhou, China
Department of Geography, Western Michigan University, Kalamazoo, Michigan, USA

L. Zhang
Key Laboratory of West China's Environmental System (Ministry of Education), College of Earth and Environmental Sciences, Lanzhou University, Lanzhou, China

ABSTRACT

Proper allocation of the limited water resources among competing uses is essential to ensure the welfare of human beings and the sustainability of ecosystems, especially in arid regions such as Northwest China. The Heihe River Watershed is the second largest inland river (terminal lake) in China, with a drainage area of 128,000 km^2. From the headwaters in the south to the lower reach in the north, the Heihe River Watershed physically consists of the Qilian Mountain, the Hexi Corridor, and the Alashan Highland. The Qilian Mountain is situated at the south of the watershed, with a peak elevation of 5,584 m. Located in the middle reach of the Heihe River Watershed, the Hexi Corridor hosts over 90% of the total agricultural oases in the watershed and supports more than 97 percent of the Heihe watershed's nearly 2 million inhibits. North of the Hexi Corridor is the Alashan Highland, an extremely dry desert with an annual precipitation below 50 mm. Since the 1970s, the increased withdrawals for agricultural irrigation in the Hexi Corridor have depleted much of the river flows to the lower reach, endangering aquatic ecosystems, accelerating desertification, intensifying water conflicts between the middle reach and lower reach users. To mitigate the water conflicts, the State Council of China has issued a "Water Allocation Plan for the Heihe Watershed Mainstream", mandating the allocation of 0.95 billion (10^9) m^3 of water annually to the lower reach under normal climatic conditions for rehabilitation of downstream ecosystems. However, are the flows from the upper and middle reaches sufficient to deliver 0.95 billion (10^9) m^3 of the water downstream annually?

This paper adapted the Distributed Large Basin Runoff Model (DLBRM) to the Heihe River Watershed to gain an understanding of the generation of glacial/snow melt, surface runoff and groundwater in the mountainous upper reach, and distribution of evapotranspiration (consumptive water use) in the middle reach of the watershed. The DLBRM was calibrated over the period of 1978–1987 (a wet hydrologic period) for each of the 9,790 cells (cell size: 4-km^2) at daily intervals. The calibration shows a 0.696 correlation between simulated and observed watershed outflows. The ratio of model to actual mean flow was 1.023. Over a separate simulation period (1990–2000, a normal hydrologic period), the model demonstrated a 0.717 correlation between simulated and observed watershed outflows, and the ratio of model to actual mean flow was 1.069. Simulation of the daily river flows for the period of 1990–2000 by the DLBRM shows that Qilian Mountain in the upper reach produced most of the runoff in the watershed. Annually, the simulated average annual flow for 1990–2000 was about 0.896×10^9 m^3 from the middle reach to the lower reach under a normal, median precipitation year (P = 50%), which falls short to meet the requirement of delivering 0.95×10^9 m^3 downstream annually mandated by the State Council 50 percent of the time. Under drier climatic conditions, even less amount of flow would be delivered downstream, posing an even greater challenge for restoring downstream ecosystem services. To tackle the increasing water conflicts among the upper, middle, and downstream users, we suggest that stakeholders from different levels of governmental agencies and private institutions be fully engaged in the watershed management process to develop a water allocation system that consists of multiple water allocation criteria, implementation plan, evaluation and feedback mechanisms.

Integrating Spatial Multi Criteria Decision Making (SMCDM) with Geographic Information Systems (GIS) for determining the most suitable areas for artificial groundwater recharge

M. Zare & M. Koch
Department of Geotechnology and Geohydraulics, University of Kassel, Kassel, Germany

ABSTRACT

The Shabestar plain is located in northwest Iran. Increasing population, agricultural development and illegal well pumping have increased the exploitation of groundwater resources in this plain. This phenomenon, along with recent droughts, has led to severe decreases of the groundwater levels and water shortages over the last years. In order to mitigate this crisis, the establishment of groundwater artificial recharge projects can be a suitable solution. An important step in the realization of such projects consists in the determination of suitable areas. In this study the Spatial Multi Criteria Decision Making (SMCDM), in conjunction with Geographic Information Systems (GIS), is used for that purpose. More specifically, seven main parameters including land slope, soil hydrologic groups, alluvium thickness, quaternary units, groundwater level are considered as the main layers in GIS, while pasture land and water drainage network are used as efficient layers for locating appropriate artificial

Table 1. Prioritizing of selected area.

Selected area No.	A (Km^2)	W(MCM)	Priority
1	21.64	3.86	3 rd
2	33.67	5.64	2 nd
3	38.63	7.65	1 st
4	5.24	1.28	7 th
5	7.25	1.74	6 th
6	9	1.91	5 th
7	8.89	2.15	4 th

recharge areas. The data layers for each one of these variables are supplied by GIS. The ranges of change of the five main layers are then classified in accordance with their importance in the locating process. These data layers are afterwards assessed with one another by means of a pairwise comparison matrix with regard to their significance to locating by applying the Analytical Hierarchy Process (AHP) technique. The selected areas are integrated with exclusionary areas from pasture lands and the presence of water drainage networks. Finally seven separate regions with an area of 124.3 km2 (=10.42% of the flood plain) are identified as appropriate flood spreading recharge areas. Based on the annual potential of runoff production, that has been calculated by Justin's method, the areas are then prioritized. Thus, region #3, with a surface of 38.6 km2, turns out to be best place for artificial recharge, as it has 7.65 million cubic meter (MCM) runoff production per year, whereas regions #2 and #1 have second and third priority, respectively.

REFERENCES

Koch, M., 2008. Challenges for future sustainable water resources management in the face of climate change, The 1st NPRU Academic Conference, Nakhon Pathom University, Thailand, October 23–24, 2008.

Patil, S.G. and Mohit, N. M., 2014. Identification of groundwater recharge potential zones for a watershed using remote sensing and GIS. International journal of geomatics and geosciences, Vol. 4, No. 3, pp. 485–493.

Figure 1. The study area and suitable areas for recharge.

Hydrological simulation in the Tiete basin

T. Rocha
School of Technology, State University of Campinas, Sao Paulo, Brazil

I.G. Hidalgo
Department of Energy System Planning, State University of Campinas, Sao Paulo, Brazil

A.F. Angelis
School of Technology, State University of Campinas, Sao Paulo, Brazil

J.E.G. Lopes
Department of Water Resources, State University of Campinas, Sao Paulo, Brazil

ABSTRACT

The operation planning of hydropower plants depends on the inflow forecasting of rivers. Usually, the inflow forecasting is done from mathematical, stochastic or hydrological models incorporated in computer systems. This paper presents the use of Soil Moisture Model Accounting Procedure (SMAP) to predict daily inflows into plants located in Brazil. SMAP (Lopes et al. 1982, 2002) is a deterministic conceptual model of the type rainfall-runoff transformation which uses the water cycle concept. The methodology consists of six steps that should be followed in order to prepare, apply, and evaluate SMAP model. They are: (1) analysis of the data consistency, (2) automatic calibration of the calculated parameters, (3) manual re-calibration of the estimated and calculated parameters, (4) adjustment of the measurement stations' weights, (5) validation/application of the model, and (6) evaluation of the results. Generalized Reduced Gradient algorithm for nonlinear optimization is employed.

The objects of study are two plants of the Tiete River basin, Ibitinga (IBI) and Bariri (BAR). SMAP is calibrated from 2003 to 2005, validated from 2005 to 2007 and applied to three periods of 2010. Its performance is analyzed through three indicators: Nash-Sutcliffe efficiency, percent bias, and ratio of the root mean square error to the standard deviation.

Figures 1–2 present a comparison between predicted and observed water inflow series for 1–15/Jan/2010. Table 1 displays the values and ratings of the performance indicators. The indicators confirm a good performance of the SMAP model with 100% of the results classified into "very good, good, and satisfactory" range, in a distributed way. The percentage relative deviations for IBI and BAR are, on average, 19% and 16%, respectively. These values are good comparing to the FIS model and to the annual reports of inflow forecasting for Tiete River basin that show a deviation between predicted and observed water inflows up to 30%. We hope to contribute for the improvement

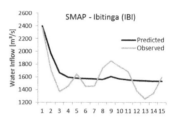

Figure 1. Predicted and observed water inflows for IBI.

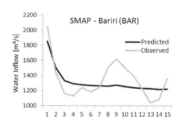

Figure 2. Predicted and observed water inflows for BAR.

Table 1. Model's performance for IBI and BAR.

Indicator	IBI Value	Rating	BAR Value	Rating
NSE	0.53	satisfactory	0.52	satisfactory
PBIAS	−3%	very good	2.4%	very good
RSR	0.68	satisfactory	0.69	satisfactory

of the power system planning, avoiding the thermal complementation and reducing the electricity cost.

REFERENCES

Lopes, J.E.G., Braga, B.P.F. and Conejo, J.G.L., 1982. SMAP – A simpliflied hydrologic model. Water Resourses Publication (Org.). Applied Modeling in Catchment Hydrology. Hittleton, 167–176.

Lopes, J.E.G., Braga, B.P.F. and Conejo, J.G.L., 2002. A simple hydrological model, applied modelling in catchment hydrology. Ed. V.P. Singh. Water Resources Publications.

Hydraulic analysis of an irrigation headworks complex in the Artibonite department (Haïti)

S. Louis & N. Gonomy
Unité de Génie Rural, Faculté d'Agronomie et de Médecine Vétérinaire, Port-au-Prince, Université d'Etat d'Haïti

N. de Ville & M.A. Verbanck
Department of Water Pollution Control (CP208), Université Libre de Bruxelles (ULB), Belgium

ABSTRACT

In the Northern part of the Republic of Haiti, water is extracted from the Artibonite River to irrigate 300 km² of a wide agricultural plain constituted primarily of paddy fields. The river comes from the Peligre reservoir where it is impounded (altitude 150 m, close to the border with Dominican Republic) to reach the Canneau irrigation dam (altitude 32 m) barring its course 70 km further downstream. The purpose of the Canneau dam is to facilitate the operation of two main irrigation channels, one which runs parallel to the river on the left side and a second one on the right side. The earth-lined Left-side Master Canal (Fig. 1), CMRG in its French denomination, is at the center of our analysis and investigative efforts.

The water intake at CMRG is controlled by the opening of a cable-hoisted radial gate (vanne-segment) fitted with counterweight (Fig. 1). The nominal design discharge sent to the CMRG is intended to be 40 m³/s.

The width of the gate is 6.25 m and its opening, from 0′ to 6′ (six inches), primarily drives the amount of irrigation water which is diverted from the river.

A first objective of the local experimental work is to calibrate the flow rate at the gate with the adequate combination of valve opening heights (creating either free or submerged orifice conditions) and the explicit account of temporal changes in the headwater tail. A second objective is to develop a 1D velocity model valid for the first two kilometers of the trapezoidal

Figure 1. Radial gate (6.25 m-width) opened at 5′6″.

Figure 2. Experimental at CMRG iron-bridge.

earth-lined CMRG canal. It is shown that there is a specific stream power range in the CMRG for which the alluvial resistance is nicely described by the integer value of control factor m = 3 (Verbanck 2008).

On the first 2 km downstream of the Canneau dam, an experimental set up (staff gauges, electromagnetic current meter and radar level gauge) is deployed, see Figure 2.

They are used for observations of water depth and exploration of the velocity field. This experimental system allowed to obtain the flow rate at Iron-Bridge and supported the elaboration of an initial abacus. It is of a diagram with three entries: flow rate, upstream head and the opening angle of the radial gate. Meanwhile, the submerged-orifice law was used to determine the effective flow coefficient (μ) of the radial gate. In addition to this, the present study aims at defining a coefficient to convert measured surface velocities into representative average 1D velocities. To achieve this, two complementary methods were used: a) exploration of the velocity field, using a portable electromagnetic current meter OTT MF Pro, b) a 1D velocity model based on the water surface slope and a bedform flow resistance term (control factor m, following Verbanck, 2008).

REFERENCES

Collins, D.L. 1977. Computation of records of streamflow at control structure U.S. Geological Survey, Water Resources Investigation Report 77–8.

Verbanck, M.A. 2008. How fast can a river flow over alluvium Journal of Hydraulic Research, Vol. 46 Extra Issue: 61–71.

Effect of emergent vegetation distribution on energy loss

E. Eriş, G. Bombar & Ü. Kavaklı
Ege University, İzmir, Turkey

ABSTRACT

For the ecological aspect, it is well known that vegetation has a fundamental function in the river environment. For fluvial hydraulics, on the other hand, presence of vegetation could be considered as a problem for decreasing the channel flow capacity due to its roughness, consequently affecting the river morphology and sediment transport (Vargas–Luna et al. 2015).

In order to understand the effects of planform areal configuration of flexible vegetation on flow resistance, a set of experiments were performed in the Hydraulics Laboratory of the Department of Civil Engineering, Ege University, in a rectangular flume which is 5 m long, 18 cm wide and 20 cm deep with transparent sides. The artificial grass representing the vegetation and its variation of front area with depth is given in Figure 1.

Experiments were conducted with two different spatial distributions of vegetation, single sided one row (grasses positioned on one half of the flume) and double sided two rows (on both sides) and preserving the total number of grasses (Fig. 2).

For each configuration, different downstream flow depths and two different discharges were applied.

Figure 2. (a) One & (b) two rows of vegetation.

Table 1. Characteristics of the experiments.

Vegetation distribution	Q l/s	h_0 cm	Fr	# of runs
One row	0.38–2.93	0.80–10.05	0.04–1.36	31
Two rows	0.29–3.06	1.00–10.05	0.03–1.30	21
Empty	0.55–2.99	1.10–10.00	0.04–1.42	23

In total 75 runs were performed with different Froude numbers as summarized in Table 1.

Energy losses were calculated using the energy balance and the continuity equations for all scenarios. The hydraulic resistance to flow through emergent vegetation was investigated in terms of coefficient of Chézy, Weisbach roughness coefficient and Manning's roughness coefficient, compared and interpreted.

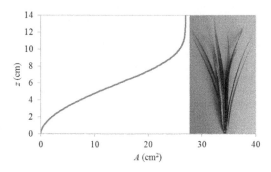

Figure 1. Relation between projected area and flow depth.

REFERENCE

Vargas-Luna A., Crosato, A. & Uijttewaal, W.S.J. 2015. Effects of vegetation on flow and sediment transport: comparative analyses and validation of predicting models. *Earth Surface Processes and Landforms*. 40(2), 157–176.

Monthly reservoir operating rules generated by implicit stochastic optimization and self-organizing maps

C.A.S. Farias, E.C.M. Machado & L.N. Brasiliano
Federal University of Campina Grande, Pombal – PB, Brazil

ABSTRACT

According to Tundisi (2008), some reasons for the so-called "water crisis" of the twenty-first century are: increasing demands for water; changes in water availability; intensification of climate extremes; and lack of articulated initiatives and government actions towards water governability and environmental sustainability.

In semiarid Brazil, the sustainable management of water resources is a complicated task because of the natural hydrological conditions (e.g. high evaporation rates, scarce and uneven space-time distribution of rainfalls and limited groundwater availability). Such environment suffers periodically from water scarcity, what compromises not only social and economic development but also water and food security (Carneiro & Farias, 2013; Farias et al., 2015).

The use of predetermined operating rules seems to be a common and practical solution to manage water systems subject to drought spells. In a water supply reservoir, for example, such rules may be derived by applying optimization procedures and linear regression equations (Loucks & Beek, 2005).

The Implicit Stochastic Optimization (ISO) procedure consists of optimizing the system operation under several inflow scenarios and using the optimal outcomes to define water allocation policies. Usually, linear regression models relate initial reservoir storages and predicted inflows to optimal releases (Farias et al., 2011; Loucks & Beek, 2005). Unlike the use of regression equations, we propose the application of a Self-Organizing Map (SOM) model as an attempt to extract possible nonlinear trends among the variables of the process.

As stated by Kohonen (1982), a SOM is a bidimensional matrix of features capable of representing multidimensional sets of data. The basic principle consists of grouping data vectors according to their similarities in a map, which may be later of use to pattern classifications and analyses.

In order to evaluate the effectiveness of the combined ISO-SOM model, we derived water allocation rules to the monthly operation of *Coremas-Mãe D'Água* water system, which is located in *Piancó* river basin, semiarid Brazil.

The outcomes showed that the ISO-SOM model is superior to the standard rules of reservoir operation. As conclusion, we believe that the combination of ISO with emerging tools such as SOM may be a promising strategy for reservoir operation.

REFERENCES

Carneiro, T. C., & Farias, C. A. S., 2013. Otimização estocástica implícita e redes neurais artificiais para auxílio na operação mensal dos reservatórios Coremas – Mãe d' Água. Revista Brasileira de Recursos Hídricos, 18(4), 115–124.

Farias, C. A. S., Bezerra, U. A., & Silva Filho, J. A., 2015. Runoff-erosion modeling at micro-watershed scale: a comparison of self-organizing maps structures. Geoenvironmental Disasters, 2(1), 2–14.

Farias, C. A. S. de, Santos, C. A. G., & Celeste, A. B., 2011. Daily reservoir operating rules by implicit stochastic optimization and artificial neural networks in a semi-arid land of Brazil. IAHS Publications, 347, 191–197.

Kohonen, T., 1982. Self-organized formation of topologically correct feature maps. Biological Cybernetics, 43(1), 59–69.

Loucks, D. P., & Beek, E. Van., 2005. Water resources systems planning and management: an introduction to methods, models and applications. Turin: United Nations Educational, Scientific and Cultural Organization.

Tundisi, J. G., 2008. Recursos hídricos no futuro: problemas e soluções. Estudos Avançados, 22(63), 7–16.

Comparison of PIV measurements and CFD simulations of the velocity field over bottom racks

Luis G. Castillo, Juan T. García, Jose M. Carrillo & Antonio Vigueras-Rodríguez
Hidr@m Group, Civil Engineering Department, Universidad Politécnica de Cartagena, Spain

ABSTRACT

A comparison of the velocity field over a bottom rack system measured by Particle Image Velocimetry (PIV) and simulated with numerical simulations (ANSYS CFX v14.0) present a good agreement (Figure 1). Laboratory measurements are taken in a physical device located in the Laboratory of Hydraulic of the Universidad Politécnica de Cartagena (Spain).

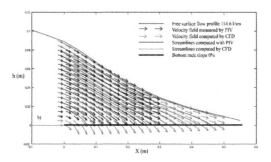

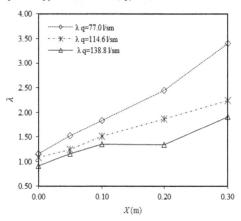

Figure 1. Velocity field and streamlines measured with PIV and simulated with CFD for rack with $m = 0.28$, horizontal slope and approximation flow, $q_1 = 114.6$ l/sm.

Figure 2. Pressure coefficient of the energy equation, λ, in cross sections located X distances from the beginning of the rack and for three specific approximation flows.

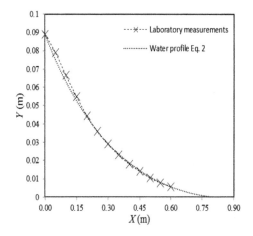

Figure 3. Comparison of the flow profile over the bottom rack from equations 1, and with laboratory measurements.

Velocity and pressure coefficients of the energy equation are obtained (Figure 2) and used to evaluate the water profile along the racks according to equation 1. Comparison with experimental data is presented (Figure 3).

$$\frac{dh}{dx} = \frac{\frac{d\lambda}{dx}h^2 + \frac{d\alpha}{dx}h(H-\lambda h) - 2C_{qh}m\sqrt{\alpha(H-\lambda h)h}}{\alpha(2H-3\lambda h)} \quad (1)$$

Pressure distribution along the flow depth is presented for several distances along the rack. Pressure results are compared with the pressure terms deviation of hydrostatic pressure proposed by Castro-Orgaz and Hager (2011).

REFERENCES

Castillo, L. & Carrillo, J.M. 2012. Numerical simulation and validation of intake systems with CFD methodology. *Proc. 2nd IAHR European Congress; Munich, 27–29 June 2012.*

Castro-Orgaz, O., W. H. Hager. 2011. "Spatially-varied open channel flow equations with vertical inertia". Journal of Hydraulic Research, 49:5, 667–675.

Noseda, G. 1956. Correnti permanenti con portata progressivamente decrescente, defluenti su griglie di fondo, L'Energia Elettrica, 565–581, 1956.

Image processing techniques for velocity estimation in highly aerated flows: Bubble Image Velocimetry vs. Optical Flow

D.B. Bung & D. Valero
Hydraulic Engineering Section, FH Aachen University of Applied Sciences, Aachen, Germany

ABSTRACT

Measuring of flow velocity in aerated flows is known to be difficult in physical models. Application of classical anemometers or ADV probes is limited to low air concentration. Thus, highly aerated flows are commonly investigated by use of intrusive needle probes (conductivity or optical fibre) which allow determination of both, air concentration and velocity, as well as related parameters (e.g. bubble chord lengths and turbulence).

In the recent past, non-intrusive image processing techniques have gained more attraction. Bung (2011), Leandro et al. (2014) and Bung & Valero (2015) demonstrate that the Bubble Image Velocimetry (BIV) technique can provide useful information on instantaneous velocity fields. This technique is based on the well-known Particle Image Velocimetry (PIV) using entrained bubbles as tracers for cross-correlation of subsequent video frames and was first introduced by Ryu et al. (2005). However, it was found that, in a quantitative view, velocities tend to be underestimated in the order of 20%.

Another method for determination of obstacle displacement or velocity, respectively, is given by Optical Flow which was introduced by Horn & Schunck (1981). This method assumes a constancy of brightness of a moving obstacle (or in the given case, a bubble) from one video frame to another. It must be noted that this method is commonly applied in the field of Computer Vision, e.g. for autonomous car-driving. Application to fluid flows is rarely found in the literature to date.

The paper compares both methods applied to aerated stepped spillway flows using a high-speed camera and highlight their capabilities and limitations. It is shown that the OF method is capable to give velocity data with the same or even higher accuracy as the BIV method. When a high sample rate is used, turbulence intensity may be determined with the OF approach and complete turbulence intensity fields can be obtained.

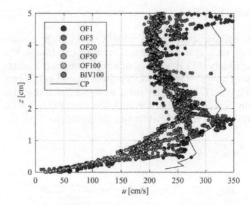

Figure 1. Velocity profiles at upstream step edge ($x = 0$ cm) from OF calculations with 1, 5, 20, 50 and 100 frame pairs compared to the BIV results with 100 frame pairs and the conductivity probe data (CP) for the high-resolution frames.

REFERENCES

Bung, D. B. 2011. Non-intrusive measuring of air-water flow properties in self-aerated stepped spillway flow. 34th IAHR World Congress, 2380–2387, Brisbane.

Bung, D. B., & Valero, D. 2015. Image processing for bubble image velocimetry in self-aerated flows. 36th IAHR World Congress, The Hague.

Horn, B. K. P., & Schunck, B. G. 1981. Determining optical flow. Artificial Intelligence, 17, 185–203.

Leandro, J., Bung, D. B., & Carvalho, R. 2014. Measuring void fraction and velocity fields of a stepped spillway for skimming flow using non-intrusive methods. Experiments in Fluids, 55 (5).

Ryu, Y., Chang, K.-A., & Lim, H.-J. 2005. Use of bubble image velocimetry for measurement of plunging wave impinging on structure and associated green water. Measurement Science and Technology, 16 (10), 1945–1953.

Biomechanical tests of aquatic plant stems: Techniques and methodology

A.M. Łoboda, Ł. Przyborowski & M. Karpiński
Institute of Geophysics Polish Academy of Sciences, Warsaw, Poland

R.J. Bialik
Institute of Geophysics Polish Academy of Sciences, Warsaw, Poland
Institute of Biochemistry and Biophysics Polish Academy of Sciences, Warsaw, Poland

ABSTRACT

Six different aquatic plant species (*Myriophyllum spicatum L., Potamogeton crispus L., Potamogeton pectinatus L., Ceratophyllum demersum L., Callitriche palustris L.* and *Elodea canadenis L.*) were collected from two lowland, sandy-bed rivers (the Wilga and the Świder) in August 2015 and in September 2015. Plants were kept in aerated water during the transportation. In the laboratory an aquarium with all equipment needed for plants to thrive was used. Next, these macrophytes were prepared in order to conduct a series of biomechanical tests (three-point bending and tension). Collected plants allowed to study biomechanical properties of broad range of aquatic macrophytes. Similar research was undertaken previously by Bociąg et al. (2009) and Miler et al. (2012).

Tests were carried out to improve knowledge of the biomechanical properties of aquatic plant stems and to show problems and difficulties in techniques and methodology of such tests which have not been pointed out in previous works. For example, aquatic plants are prone to dry very quickly when they are outside their natural environment. These tests were made in dry and wet conditions with a benchtop materials testing machine. The diameter and cross-section of the stem, sample length and gauge length were measured for each sample.

Preliminary results showed that there is significant difference between results obtained from dry and from wet tests. These differences were not equal between species nor between bending and tension tests. Pieces of *Potamogeton pecinatus L.* were most resistant to stretching in tension tests, while *Myriophillum spicatum L.* had highest strain values. In bending tests, dry specimens of *Myriophillum spicatum L.* had two times lower values of measured force required to bend stem than in wet conditions.

The conducted trial tests show that the diameter of the stem, its internal structure and the growth stages of plants are of great significance regarding the value of the measured forces. The obtained results could be crucial in identifying a relationship between plant biomechanics and flow resistance. Significant problems, such as choosing the right way to protect the ends of the sample from damage in the clamps of the device used and minimizing the time after removal of the sample from water for measurement are highlighted and discussed briefly. Accurate scheme of additional scaffolding was attached, as well as photographs of a prepared specimen, a fixed between clamps specimen, size and external morphology comparison between collected species and a photography showing how stem cross-section shrink when it was drying in the laboratory. Brief recommendations were added to point out issues that should be taken into consideration in following experiments on aquatic plants.

REFERENCES

Bociąg, K., Gałka, A., Łazarewicz, T. & Szmeja, J. 2009. Mechanical Strength of Stems in Aquatic Macrophytes. *Acta Societatis Botanicarum Polonise* 78(3): 181–187.

Miler, O., Nikora, V., Albayrak, I. & O'Hare, M. 2012. Biomechanical properties of aquatic plants and their effects on plant-flow interactions in streams and rivers. *Aquatic Sciences* 74(1): 31–44.

LS-PIV procedure applied to a plunging water jet issuing from an overflow nappe

Y. Bercovitz, F. Lebert & M. Jodeau
EDF R&D National Hydraulic and Environment Laboratory (LNHE), Chatou, France

C. Buvat & D. Violeau
EDF R&D National Hydraulic and Environment Laboratory (LNHE), Chatou, France
Saint-Venant Laboratory for Hydraulics, Chatou, France

L. Pelaprat
Ecole Nationale des Travaux Publics d'Etat, Villeurbane, France

A. Hajczak
Ecole Nationale Supérieure de Mécanique et d'Aérotechnique, Poitier, France

1 INTRODUCTION

The accumulation of statistical data and scientific advances in hydrology gives rise to re-evaluations of the extreme flood discharges that can transit hydroelectric facilities. Moreover, regulations are evolving to ensure greater control over safety issues, an example of which could be an increase in the number and frequency of extreme events taken into account for the design of hydroelectric projects. It is therefore important to gain a better understanding and knowledge of the physical processes necessary for the design of protective measures.

2 CONTENT

The aim of this work is to understand the behaviour of a sheet of water falling freely under gravity, subject to instability and breaking up into droplets. Such behaviour significantly affects the pressure force at the toe of large dams equipped with ogee-type weirs. In this laboratory experiment, the LS-PIV (Large Scale – PIV) procedure is applied to determine the surface velocity field of a water jet issuing from a sharp-crested weir. The fall is 5 meters high and the weir is 40 cm wide with a maximum discharge flow of $0.050\,m^3/s$. In order to obtain high resolution measurements, the water jet was analysed using a high speed video camera (1000 fps). The initial results have been post-processed with the FUDAA-LSPIV software, co-developed by EDF and IRSTEA. The patterns of deformation of the water-air interface were recorded to determine the velocity vectors. The deformation patterns are relatively small thanks to the high frequency acquisition of the video camera. No special tracers were required as the high frequency acquisition ensures a good correlation factor is consistently obtained. This method allows for continuous monitoring of a large velocity field at a high time frequency.

The discussion covers details of the corrections that were made in order to integrate the nappe trajectory and presents the finding of a characteristic distance following which energy dissipation becomes significant.

These results indicate a close fit between the measured velocities and the velocity profile of gravity drop flows up to a point that corresponds to a fall of about 3 m, beyond which the measured velocities drop off. Visual observations reveal that the jet broadens at this distance, which explains this sudden, rapid deceleration. Thus, without having observed, in the strictest sense of the word, the break-up we demonstrate empirically that there is a characteristic length beyond which the velocity gradient diminishes.

In the laboratory experiments, this characteristic length is much greater than the breakup length obtained using the Castillo (2006) equation. A reasonable explanation has not as yet been found for these differences.

REFERENCE

Castillo L. 2006. Aerated jets and pressure fluctuation in plunge pools. *Proc. Int. Conf. The 7th International Conference on Hydroscience and Engineering*, Philadelphia, 1-23, M. Piasecki and College of Engineering, Drexel University, USA.

PIV-PLIF characterization of density currents

B. Pérez-Díaz
Environmental Hydraulics Institute "IH Cantabria", Universidad de Cantabria, Spain

P. Palomar
Environmental Hydraulics Institute "IH Cantabria", Universidad de Cantabria, Spain
Ministry of Agriculture, Food and Environment, Spain

S. Castanedo
Environmental Hydraulics Institute "IH Cantabria", Universidad de Cantabria, Spain
Departamento de Ciencias y Técnicas del Agua y del Medio Ambiente, Universidad de Cantabria, Spain

ABSTRACT

Saline density currents are horizontal flows driven by the density difference between the environmental fluid and the density current. In recent years, as the problems related to environmental conservation and coastal development have become more serious and complicated, it has become important to understand the behaviour of these flows. They are common in nature, e.g. in the far field region of hypersaline discharges from desalination plants (Palomar & Losada, 2010).

This work carries out an experimental characterization of saline density currents through advanced non-intrusive laser optical techniques PIV (Particle Image Velocimetry) and PLIF (Planar Laser Induced Fluorescence). By means of synchronized PIV-PLIF techniques, high-quality accurate instantaneous measurements of velocity and concentration are obtained (see Figure 1). While other laboratory tests have focussed on the dynamics of the current head and current spreading, the aim of these experiments is to study the quasi-steady flow properties of the current body generated by a constant flux release.

Seeking elucidation about the most influential variables in the behaviour of density currents, different experimental set-up varying the initial conditions (see Table 1) were carried out in a $3 \times 3 \times 1$ m tank in the Environmental Hydraulics Institute of Cantabria. Through PIV-PLIF result analysis, important conclusions about the influence of these variables on the mixing at the interface between fluids have been obtained.

Table 1. Configurations tested to characterize density currents.

Cases → Properties ↓		C1	C2	C3	C4	C5	C6
Slot dim. (m)	b_o	0.100	0.100	0.100	0.100	0.100	0.100
	h_o	0.025	0.015	0.025	0.025	0.025	0.025
Water depth, H_a (m)		0.460	0.460	0.460	0.420	0.360	0.460
Slope angle, α (%)		1.00	1.00	1.00	2.50	4.50	1.00
Density difference, $(\rho_a - \rho_o)$ (kg/m³)		3.145	3.100	3.130	3.070	3.140	11.08
Discharge flow rate, Q_o (l/min)		14.60	15.10	19.20	14.98	14.09	14.89
Discharge velocity, U_o (m/s)		0.095	0.164	0.125	0.098	0.098	0.097

To carry out a quantitative comparison between currents, the stable mixing rate along each current is evaluated, which is commonly known as Entrainment (E) in scientific literature (Morton et al. 1956; Ellison & Turner 1959). As an example, keeping constant the rest of variables, steeper slopes and higher flow rates favour mixing, reaching stable mixing rates (E) values two times higher than values obtained in the corresponding base case ($E \sim 2 \cdot 10^{-2}$).

In addition, a high resolution and quality experimental database has been generated, which will allow to calibrate/validate hydrodynamic modelling tools.

REFERENCES

Ellison, T. H., & J. S. Turner, 1959, Turbulent entrainment in strati_ed ows, Journal of Fluid Mechanics, 6 (03), 423, doi:10.1017/S0022112059000738.

Morton, B. R., G. Taylor & J. S. Turner. 1956, Turbulent Gravitational Convection from Maintained and Instantaneous Sources, Proceedings of the Royal Society A: Mathematical, Physical and Engineering Sciences, (1196), 1–23, doi:10.1098/rspa.1956.0011.

Palomar, P., & I. J. Losada, 2010, Desalination in Spain: Recent developments and recommendations, Desalination, 255 (1–3), 97–106, doi:10.1016/j.desal.2010.01.008.

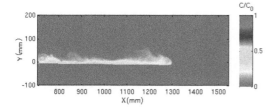

Figure 1. Snapshot normalised concentration image.

Physical model tests for the construction stages of large breakwaters. Case studies to the ports of Barcelona and Cruña (Spain)

R.M. Gutierrez, J. Lozano & J.M. Valdés
Centro de Estudios y Experimentación de Obras Públicas (CEDEX), Spain

ABSTRACT

Physical models tests are extensively employed, for testing and optimization of final breakwaters de-signs. However, their application to construction stages analysis of these structures is less common and, therefore, less well known, but their use facili-tates the design of such stages and the construction safety in general and especially that of the people in-volved in it.

Given this situation, the present paper provides an overview of several possibilities offered by physical experimentation on breakwater construction and presents two examples conducted at the Center of Ports and Coastal Studies of CEDEX: Barcelona port enlargement breakwater and Coruña Outer Port breakwater (Figure 1), which are representative of construction stage tests of the two most common breakwaters types: vertical and rubble mound.

The conclusion of the two cases described in the full paper and from others construction stages tests conducted in the CEDEX, can be summarized as follows:

– Usefulness of physical models as a tool for studying the construction phases of port breakwaters
– Application to both rubble mound and vertical breakwaters
– Possibility to analyze the construction phases of the type section (2D tests) as well as the construction progress as a whole (3D test)
– In both cases the size of the models must be as large as possible, especially for the rubble mound breakwaters, considering that among its components flow should be turbulent, as actually occurs in the prototype, for which Reynolds number should be higher than 30,000.

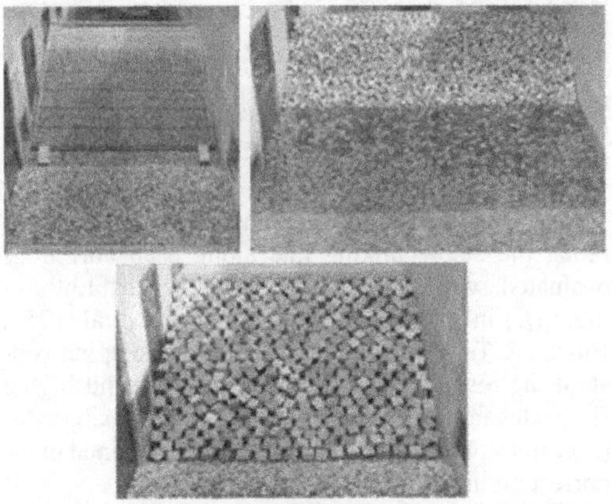

ACKNOWLEDGMENTS

Thanks are due to Spanish State Port Public Body and Port Authorities of Barcelona and Coruña for entrusting CEDEX to perform these tests, which can be extended to many other marine structures.

Figure 1. 4 Coruna outer port breakwater – Construction stages 1, 4 and 7.

Low-cost 3D mapping of turbulent flow surfaces

A. Nichols & M. Rubinato
Civil and Structural Engineering Department, The University of Sheffield, UK

1 INTRODUCTION

Recent work has shown that water surface fluctuations can be associated with the underlying velocity field and turbulence, which could in turn be related to the flow conditions, boundary conditions and hydraulic processes.

This work presents an innovative low-cost method to measure surface water patterns by using a Microsoft Kinect sensor, which has the ability to measure at reasonable spatial resolutions and in three dimensions. Kinect sensors have been used to measure fluid surfaces in the form of gravity waves generated in coloured (opaque) water and sand flows. They have never been used to measure turbulence-generated water surface roughness. They have also never been used to measure clear water surfaces.

Here, a Kinect sensor is positioned above an experimental channel with a gravel bed within the University of Sheffield hydraulics laboratory. Initially, large-scale gravity waves are generated using a manual paddle for a range of clear flow conditions and the resulting fluctuations in bed measurements are validated against data collected with a conductance-based wave probe in the same location. Subsequently, the smaller scale turbulence-generated roughness is measured by the Kinect sensor and the wave probe. The hypothesis is that the local surface gradient refracts the infra-red light, generating measurable fluctuations in the Kinect signal.

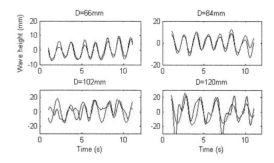

Figure 1. Gravity waves (Blue – wave probe; Black – Kinect).

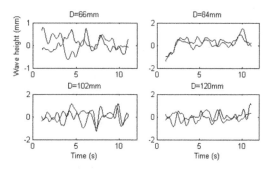

Figure 2. Flow waves (Blue – wave probe; Black – Kinect).

2 RESULTS & CONCLUSIONS

Figure 1 shows gravity waves generated in a sub-set of four shallow flows, measured by conductance-based wave probe and by the Kinect sensor. Figure 2 shows turbulence-generated flow surface waves. The data show that the Kinect sensor can measure gravity waves in clear water. For flow surface waves there is ambiguity in individual features, but some agreement for certain flows. It is postulated that a flat flume bed would remove the ambiguity, and that with further development the sensor may have the potential to measure small scale turbulence-generated water surface roughness.

Links are also demonstrated between the free surface fluctuation and the variability in the flow velocity measured via Acoustic Doppler Velocimetry, suggesting that the remotely sensed surface data can be used to infer turbulence processes.

These data provide the first step in a non-intrusive system for characterising turbulence driven processes in shallow turbulent flows. The clear-water Kinect technique can also allow simultaneous optical techniques such as PIV & LIF, and may enable surface measurement in clear streams or clear laboratory flows. It is hoped that this work will stimulate deeper study and accelerated development of the techniques proposed here.

Sediment transport measurements in the Schelde-estuary: How do acoustic backscatter, optical transmission and direct sampling compare?

S. Thant
Antea Group, Antwerp, Belgium

Y. Plancke, S. Claeys
Flanders Hydraulics Research, Antwerp, Belgium

ABSTRACT

Measuring sediment transport remains one of the most challenging aspects in river engineering. In 2014, Flanders Hydraulics Research performed several field measurement campaigns in the Schelde-estuary. The goal of these campaigns was to gather datasets for numerical model (Delft3D and Telemac-suites) validation of sediment transport and to increase our system understanding. During the measurements campaigns, several problems (e.g. bio-fouling) occurred while using optical backscatter (OBS) devices over longer periods (~4 weeks). Where simultaneous measurements of flow characteristics using Acoustic Doppler Profiler were performed, acoustic backscatter data was used to make estimates of the sediment concentration after the bio-fouling occurred on the OBS-sensors. Unfortunately, no specific field campaign was organized to calibrate the acoustic backscatter, so the results had large uncertainties.

In 2015 a specific field campaign was organized to calibrate both the acoustic backscatter signal (Nortek AquaDopp) and the optical transmission signal (LISST-100X), while collecting multiple water and sediment samples using pump samplers and a water trap.

Results show a good agreement between volume concentrations of the LISST-100X compared to sediment concentrations determined by filtration of the pump samples and Water Trap samples. Backscatter data (AquaDopp) confirms the sediment concentrations obtained by the pump samples, justifying the general use of backscatter data as a proxy for sediment concentration.

Correlations between pump samples (Oosterweel and Ketelplaat) and LISST-100X and backscatter data were moderate, becoming stronger when making a distinction in measurement location. Distinctions in sediment type (sand/silt) gave moderate to weak correlations.

The low correlation between the pump samples and the indirect measurements can be due to the altering of the sediment samples during effective sampling, transport and handling of the samples in the lab (e.g. destruction of flocs, formation of new flocs, ...). Furthermore, measuring methods of the indirect techniques (LISST-100X, AquaDopp) are assumed to be suitable for the terrain conditions present. However, because of the high variability in transported sediment (sand, silt, flocs, shape, ...), both the optical and acoustic measuring technique will have limitations, ensuring indirect techniques to be an estimation of the real sediment concentration.

Median grain sizes (d50) measured by the LISST-100X are significantly higher for all locations than the corresponding d50 values obtained by analysis of the water samples (pump samples, water trap samples, Delft bottle samples). This deviation can originate from the initial presence of sediment flocs in the system, which are broken up by pumping up the water, during transport or by handling of the water samples in the lab, leading to a higher concentration of fine grained particles and lower median grain sizes.

Lab conditioned measurements by the LISST-100X on calibrated sand (105 μm) also showed an overestimation of the expected median grain size (130 μm–150 μm), it is not clear how much of the deviation in d50 between LISST-100X and water samples is due to measurement configuration or effective change in sediment properties.

Field-deployable particle image velocimetry with a consumer-grade digital camera applicable for shallow flows

K. Koca, A. Lorke & C. Noss
Institute for Environmental Sciences, University of Koblenz-Landau, Germany

ABSTRACT

Detailed flow structure measurements in natural flows can provide new insights into the understanding of environmental fluid mechanics processes. Inspired by the need for flow detailization in natural flows, a number of researchers (Liao et al., 2009; among others) employed the well-established Particle Image Velocimetry (PIV) technique, in field-based studies. PIV is advantageous over typical measurement techniques (e.g. Acoustic Doppler Velocimetry) because it provides direct measurements of instantaneous velocity field, its spatial derivatives and spatial covariances in two (or three) spatial dimensions. These features allow for the calculation of instantaneous vorticity, dissipation rates (Westerweel et al., 2013), and turbulent wave number spectra without relying on Taylor's frozen turbulence approximation, as well as observation of coherent flow structures. The field-PIV systems developed so far, however, utilized high power, sophisticated laser systems, as well as sophisticated cameras, which makes them expensive and requires extensive deployment effort. We describe an alternative and inexpensive field-deployable PIV system based on a consumer-grade camera (GoPro Hero 4, GoPro Inc., USA) and a 225 mW (532 nm), continuous-wave laser module (Hercules, LaserGlow, Canada), which can be deployed in very shallow flows with a minimum water depth of 6 cm. To validate the developed system, simultaneous velocity measurements were performed in a flume using a Vectrino Profiler (Nortek, AS). The flow depth was constant at 30 cm, while the mean flow velocity, U, varied between 1.5 cm s^{-1} and 37.3 cm s^{-1} (four runs). Good agreement was found in a direct comparison of velocity time series (Figure 1). The velocities measured by the PIV were observed to be slightly less than those measured by Vectrino Profiler. The differences between the mean longitudinal velocities varied between 0.2% and 6.9%, with the biggest difference observed for $U = 37.3$ cm s^{-1}, whereas the differences were between 0.7% and 5.5% for the mean vertical velocities.

The root-mean-square velocity fluctuations measured by both instruments were in a reasonable agreement, with differences between 5.6% and 8.2% for the

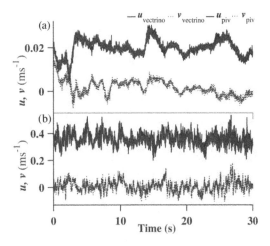

Figure 1. Comparison of velocity time series obtained by two instruments. a) $U = 1.5$ cm s^{-1}, b) $U = 37.3$ cm s^{-1}. Solid lines denote longitudinal velocity, u, while dashed lines denote vertical velocity, v.

longitudinal component, and between 5% and 13.9% for the vertical component. Despite the limitations due to the use of continuous illumination as opposed to the short duration, high intensity pulses in commercial PIV systems, small sensor with compressed video output, and lack of shutter speed control, we demonstrated that the proposed PIV system shows comparable performance to Vectrino Profiler, and can be practically used in field-based research, particularly in challenging shallow and slow flow applications.

REFERENCES

Liao, Q., Bootsma, H. A., Xiao, J., Klump, J. V., Hume, A., Long, M. H., and Berg, P. 2009 Development of an in situ underwater particle image velocimetry (UWPIV) system. Limnol. Oceanogr. Methods, 7, 169–184.

Westerweel, J., Elsinga, G. E., Adrian, R. J. 2013. Particle Image Velocimetry for complex and turbulent flows. Annual Review of Fluid Mechanics, 45, 409–436.

Numerical modelling of meandering jets in shallow rectangular reservoir using two different turbulent closures

Y. Peltier
University of Liege, ArGEnCo Department, Research Group of Hydraulics in Environmental and Civil Engineering, Liege, Belgium
Université Paris-Est, Laboratory for Hydraulics Saint-Venant (ENPC, EDF R&D, CEREMA), Chatou, France

S. Erpicum, P. Archambeau, M. Pirotton & B. Dewals
University of Liege, ArGEnCo Department, Research Group of Hydraulics in Environmental and Civil Engineering, Liege, Belgium

ABSTRACT

Natural or engineered reservoirs are very common structures in hydraulic engineering. They are used as storage reservoirs for flood management or as settling reservoirs for trapping sediments and/or pollutants. The control of the sediment transport within these structures is vital to achieve their cost-effective and sustainable management.

This control can only be achieved through the robust modelling of the flow patterns and of their interaction with the sediments during the design process. Indeed, standard design approaches merely based on the reservoir volume have shown their limitations (Dufresne et al. 2012). The flow patterns must be taken into account for optimally designing the reservoir (Peltier et al. 2014a, 2014b), as complex flow fields develop in such structures (large scale turbulent vortices, meandering jets), even for simple geometries (Peltier et al. 2014b). In rectangular shallow reservoirs, depending on the geometrical parameters and on the hydraulic conditions, four distinct flow regimes can thus be defined (symmetric, asymmetric, meandering and unstable).

In a recent study Peltier et al. (2015) confirmed that the shallow water equations coupled to a k-ε model are able to model flows in rectangular reservoir even for a complex flow regime like the meandering one. Moreover, by just adjusting the roughness height in the friction modelling they showed that the results could be improved.

In the present work, the influence of the turbulent closure on the modelling of two meandering flows is analyzed. The first turbulent closure accounts for the 2D horizontal turbulent mixing (k-ε), while the 3D vertical turbulent mixing is treated with an algebraic model (Elder) and three different tuning parameters. The second turbulent closure uses a subgrid-scale model for modelling the unresolved scales (small eddies) of the shallow water equations (Smagorinsky model). Simulations are compared to experiments

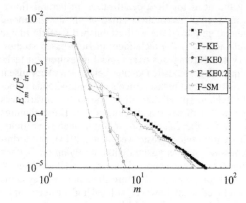

Figure 1. Mean fluctuating kinetic energy contained in the mth modes and normalized by the square of the velocity at the inlet (KE: k-ε; SM: Smagorinsky).

using flow descriptors obtained by the Proper Orthogonal Decomposition of the fluctuating velocity fields (e.g. the eigenvalues = the energy in Figure 1). Results indicate a great sensitivity to the 3D turbulence modeling.

REFERENCES

Dufresne, M., Dewals, B., Erpicum, S., Archambeau, P., Pirotton, M., 2012. Flow patterns and sediment deposition in rectangular shallow reservoirs. Water Environmental Journal, 26(4): 504–510.

Peltier, Y., Erpicum, S., Archambeau, P., Pirotton, M., and Dewals, B., 2014a. Experimental investigation of meandering jets in shallow reservoir. Environmental Fluid Mechanics, 14: 699–710.

Peltier Y., Erpicum S., Archambeau P., Pirotton M. and Dewals B., 2014b. Meandering jets in shallow rectangular reservoirs: POD analysis and identification of coherent structures. Experiments in Fluids, 55(6): 1–16.

Peltier, Y., Erpicum, S., Archambeau, P., Pirotton, M., Dewals, B., 2015. Can Meandering Flows in Shallow Rectangular Reservoir Be Modelled with the 2D Shallow Water Equations? Journal of Hydraulic Engineering 141, 04015008.

Turbulent momentum exchange over a natural gravel bed

C. Miozzo & A. Marion
Dipartimento di Ingegneria Civile, Edile ed Ambientale, Università degli Studi di Padova, Padova, Italy

A. Nichols & S.J. Tait
Department of Structural Engineering, University of Sheffield, Sheffield, UK

ABSTRACT

Exchange processes occurring between a porous bed and the stream flow above play an important role in controlling the transport of contaminants and other substances in rivers and streams. Zhou and Mendoza (1993) theorized that the migration of solutes between streams and flat gravel beds was due to turbulent momentum transfer across the stream–subsurface interface (Packman et al., 2010).

A series of experiments were carried out in a tilting rectangular laboratory flume in order to investigate the momentum transfer between a turbulent flow and a porous gravel bed of uniform thickness. In particular, the goal of this study was to quantify and compare the spatial pattern of the momentum flux with the spatial pattern of the bed, and to calculate the scale of the interactions. Three different flow conditions were considered, which were characterised by distinct flow depths, and were designed to avoid bed particle entrainment (Shields number <0.04).

Boundary shear stress is a measure of the average momentum removed from a moving fluid by a boundary (Cooper & Tait, 2010). PIV measurements were used to calculate the boundary shear stress, to define the different roles of the bed shape and of the flow conditions in controlling the momentum flux.

The cross-correlation functions between the near-bed momentum flux and the physical bed profile were characterised by the same periods independently from the flow depth, suggesting that the flow conditions did not significantly affect the organisation of the momentum flux. The disorganised gravel bed shape contains more than one dominant length-scale, but there is not a dominant spatial period coincident with the diameter of the bed particles (d_{50}). It is of note that the spectra of the of the cross correlation of the momentum flux with the bed profile (Figure 1) have the same periodicity that is presented by the bed structure (Figure 2). Therefore, it is suggested that the shape of the bed and the individual particles control the momentum transfer between the bed and the overlying fluid rather than the flow conditions. It was also found that the dominant

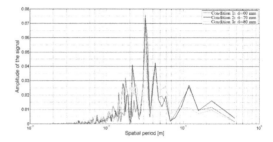

Figure 1. Spectral analysis for the three flow conditions.

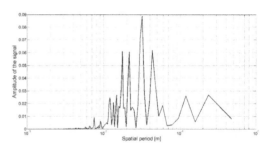

Figure 2. Spectral analysis of the gravel bed profile.

length scale in the momentum flux may be governed by flow cells within the gravel bed, and thereby relate to the bed thickness.

REFERENCES

Cooper, J.R. & Tait, S.J. 2010. Examining the physical components of boundary shear stress for water-worked gravel deposits, *Earth Surface Processes and Landforms*, 35(10), 1240–1246.

Packman, A., Salehin, M., and Zaramella, M., 2004. Hyporheic Exchange with Gravel Beds: Basic Hydrodynamic Interactions and Bedform-Induced Advective Flows. *J. Hydraul. Eng.*, ASCE, 130(7), 647–656.

Zhou, D. and Mendoza, C., 1993. Flow through Porous Bed of Turbulent Stream. *J. Eng. Mech.*, ASCE, 119(2), 365–383.

Water as a renewable energy

Planning a small hydro-power plant in Lübeck (Germany) – Who owns the water?

M. Oertel
*Hydraulic Engineering Section, Civil Engineering Department, Lübeck University of Applied Sciences
Lübeck, Germany*

ABSTRACT

Nowadays, renewable energy is essential to reduce CO_2 emission and to replace non-eco-friendly power plants. Therefore, also small hydro-power plants can be constructed at flow systems with small discharges. The city of Lübeck is located near the Baltic Sea in the Northern part of Germany. The river Trave surrounds the historic city center. Several hundred years ago another river (Wakenitz) was directly connected with the Trave flow system. But with beginning of the 12th century a continuous anthropogenic adaption of both river systems lead to the present situation, which also include a national waterway. Trave and Wakenitz are arranged according to a water level difference of approximately 3.5 m. Via a culvert connection system several hundreds of liters per second are continuously available. Hence, a hydro-power plant was setup which produces small amounts of annual energy.

Due to the European Water Framework Directive (EU-WFD) both river systems had to be connected with a fish-friendly open channel. But in consequence, less discharge could be guaranteed for the existing hydro-power plant. Hence, a new small hydro-power plant was investigated – in form of the Archimedes screw – to use necessary dotation discharges to generate energy. A hydraulic calculation results in an installed 100 kW screw turbine with an averaged annual energy production of approximately 310,000 kWh. Including investigation and operation costs amortization can be assumed to be reached after 13-18 years.

Nevertheless, a large area of conflict has been identified: who owns the water? A local energy provider operates the existing hydro-power plant and the water passes a connected lake system, thus the water quality might be affected. By changing water amounts for the present flow system current water contract situations will change, which lead to necessary round table discussions to find adequate water demands and distributions. All involved parties need to discuss advantages and disadvantages and find open-minded new solutions to satisfy all requirements.

Consequently, also small hydro-power projects can be performed successfully, but without having the charm to achieve high profit results. These structures are more or less ideology projects, but with a realistic chance to receive small amounts of cash valued output. Concluding, good data availability combined with conservative planning processes can be used for serious small hydro-power plant calculations.

Effects of changes in flow velocity on the phytobenthic biofilm below a small scale low head hydropower scheme

L. O'Keefe & S. Ilic
Lancaster Environment Centre, Lancaster University, Lancaster, UK

ABSTRACT

This study presents a spatial analysis of physical and biotic river conditions below a low head hydro scheme in the River Goyt, UK. The overall aim was to assess whether changes in localized flow velocity, introduced by low head hydro, affect phytobenthic biofilm biomass.

The area below Stockport Hydro was surveyed during low flow periods to cover as much of the localized area as possible. The study was split over two days to in a bid to collect single point measurements and profile measurements respectively. During the first field campaign single point near bed flow velocity was measured using a Valeport electromagnetic flow meter, depth was measured using a simple rigid meter rule and elevation was measured using a Trimble RTK GPS. During the second field campaign velocity profiles where measured at the near and far side river bank (hydro side and non-hydro side) using a Sontek Acoustic Doppler Current Profiler (ADCP). The ADCP was anchored in stationary positions for 5 minute time periods. On both occasions phytobenthic biomass was measured using an in situ fluorometry device.

Single point and profile measurements of flow velocity and velocity vectors, elevation, depth and biofilm biomass were interpolated and mapped using kriging methods in Surfer Software. The relationship between biofilm biomass and flow velocity was explored using Linear Regression.

Results showed evidence of high flow velocity on the hydro side of the river and low flow velocity on the non-hydro side of the river. This matched findings by the Environment Agency at Romney Weir in the River Thames (Mould *et al.*, 2015). A number of distinct hydrological and morphological features, typically associated with the interface of two flows (Szuipany *et al.*, 2009), were defined at the point where the outlet discharged water into the main river channel. Biofilm biomass appeared lower on the hydro side of the channel but no obvious relationship with flow velocity was observed. Future analysis will include evaluation of phytobenthic species on either side of the river channel and an investigation into the combined effects of a number of variables on phytobenthic biomass.

REFERENCES

Mould, D., Whitmore, J., Bentley, S., Thomas, L. & Moggridge, H. 2015. Assessment of the impact of hydropower on weir pool features. Environment Agency Report – SC120077/R1. Available at https://www.gov.uk/government/publications/hydropower-assessment-of-the-impact-on-weir-pool-habitats [accessed 18th February 2016].

Szupiany, R.N., Amsler, M.L., Parsons, D.R. & Best, J.L. 2009. Morphology, flow structure, and suspended bed sediment transport at two large braid-bar confluences. Water Resources Research 45 (5).

New concepts in small hydropower plants schemes in Romania

F. Popa
ISPH Project Development, Bucharest, Romania

D. Florescu
Individual Hydropower Consultant, Ploiesti, Romania

B. Popa
University Politehnica of Bucharest, Bucharest, Romania

ABSTRACT

In Romania, at the end of the communist era, the hydropower schemes were over dimensioned in terms of installed capacity, hence in hydraulic structures and equipment as well. Today, more than 100 small hydropower plants (hydropower plants with installed capacity less than or equal to 10 MW) must be refurbished until the end of 2016 in order to benefit from the E-RES support system.

According to the existing regulations, the refurbished SHPPs receive 2 green certificates per MWh produced, which can be sold for more than 27 Euros each. This is the only chance for these plants to become profitable. The existing hydropower schemes must be analysed so as to determine whether they were adequately dimensioned and whether the river sector was fully developed. In order to use the existing assets as much as possible, to fully develop the partially developed river sectors and to correct the deliberate designing errors, new concepts must be applied for the refurbishment stage.

A bibliographical survey is presented regarding: the SHP potential and the possible further development of SHPPs in different countries, policies, electricity markets, innovative concepts and technologies, optimal choice of location, sizing, choice of equipments and operation, costs involved by a refurbishment or construction of a new plant.

It is presented the case of Romania and the capitalization program for the small hydropower potential initiated at the beginning of '80s, unfortunately, having the same centralized character as the political system existing those years.

Because of this, the establishment and the parameterization of the future SHPPs locations, the defining of the arrangement solutions as well as the design of the constructions and electromechanical equipments, suffered from those constraints imposed by the requests for the mandatory standardization in the most limited number of options.

Without going into statistical details, we need to underline the fact that the electrical power production obtained in the almost 200 SHPP built in this program achieved, on the average, less than 70% of its project value during the first years of operation and followed a constantly descending trend afterwards. The analysis of each SHPP shows significant differences, from 100% achievement of project electricity production to 20%, which results in the decommissioning of the plant in a very short time.

The paper presents the constraints and highlights new concepts and technical solutions intended to improve the capitalization of the useful potential. Some successful case studies were presented demonstrating that it is possible to accomplish a fully hydropower developed river by correcting all the faults in conception, design and constructions by applying new and appropriate concepts.

REFERENCES

Colesca, S.E. & Ciocoiu, C.N. 2013. An overview of the Romanian renewable energy sector. *Renewable and Sustainable Energy Reviews*, 24: 149–158.

Kusakana, K. 2014. A survey of innovative technologies increasing the viability of micro-hydropower as a cost effective rural electrification option in South Africa. *Renewable and Sustainable Energy Reviews*, 37: 370–379.

Mishra, S., Singal, S.K. and Khatod, D.K., 2011. Optimal installation of small hydropower plant – A review. Renewable and Sustainable Energy Reviews, 15, 3862–3869.

Okot, D.K. 2013. Review of small hydropower technology. *Renewable and Sustainable Energy Reviews*, 26: 515–520.

Rahi, O.P. & Chandel, A.K. 2015. Refurbishment and uprating of hydropower plants—A literature review. *Renewable and Sustainable Energy Reviews*, 48: 726–737.

Sachdev, H.S., Akella, A.K., Kumar, N. 2015. Analysis and evaluation of small hydropower plants: A bibliographical survey. *Renewable and Sustainable Energy Reviews*, 51: 1013–1022.

Sternberg, R., 2010. Hydropower's future, the environment, and global electricity systems. Renewable and Sustainable Energy Reviews, 14, 713–723.

Fish behavioral and mortality study at intake and turbine

P. Rutschmann & S. Schäfer
Hydraulic and Water Research Engineering, Technische Universität München, Germany

B. Rutschmann
HOBOS Team, University of Würzburg, Germany

F. Geiger
Hydraulic and Water Research Engineering, Technische Universität München, Germany

ABSTRACT

Over the last three years tests with almost 2000 fish have been conducted at the patented TUM Hydro Shaft Powerplant (TUM-HSPP). The tests involved five different species, namely trouts, graylings, barbels, bullheads and minnows with sizes between 50 and 200 mm. The turbine mortality very much differed among the tested fish species, and a large variation is detected over the various test runs. The contribution analyses the results species independently and shows that the trends to damage depends on the length of the fish, the intake flow velocities and the operational regulations of the turbine. The present paper discusses the various influences of the hydraulic conditions at the trash rack and the operational setup of the turbine and tries to summarize these effects. Also the paper tries to separate the size dependent probability of trash rack passage from that of the pure turbine damage.

The tests showed that the fish in general tried to avoid entrance through the horizontal and river bed parallel trash rack into the turbine due to the unfamiliar flow situation. Therefore, the horizontal trash rack very much acted as a behavioral barrier. Underwater cameras allowed to observe that fish had no difficulty to swim over the intake at the trash rack for several hours. Certainly the low velocities in the intake section were the reason for this.

The paper shows that the mortality at a hydropower plant depends on the probability of trash rack passage and the probability of lethal damage in the turbine. These two probabilities have been evaluated for the TUM-HSPP. The probability of bypass passage at the TUM-HSPP design depends on the fish length and can be described by a power law function for a given intake velocity. Increasing intake velocity result in a lower probability of bypass use.

The damage probability in the turbine depending on the fish length is reproduced by a linear function as described in literature (e.g. Monten, 1985). By trend throttling of the turbine and therefore reduction of

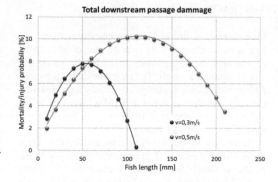

Figure 1. Overall mortality for varying fish lengths.

flow through the turbine has a negative effect on fish mortality.

The overall mortality at the TUM-HSPP results therefore from the two effects described above, i.e. from the combined probabilities from trash rack passage and mortality at the turbine. The combination of the two result in the figure shown below (Figure 1). It shows that the mortality reaches a maximum for an intermediate fish length and drops for smaller and larger fish respectively. The average damage over the range of fish sizes being able to enter the trash rack can be determined through integration of the mortality curve. For the small and rapidly spinning Lab turbine it computes to 7.2% for the higher velocity (0.5 m/s) and to 4.5% for the lower velocity (0.3 m/s). In case of a 400 kW powerplant these values were reduced by at least a factor of 2. Individual fish longer than 200 mm are completely protected in the TUM-HSPP.

REFERENCE

Monten E. (1985). Fish and Turbine – Fish Injuries During Passage Through Power Station Turbines, Vattenfall, Stockholm, ISBN 91-7186-247-1.

Long-term evaluation of the wave climate and energy potential in the Mediterranean Sea

George Lavidas
Institute for Energy Systems, University of Edinburgh, Scotland, UK
CERES Global, European Offices, Brussels, Belgium

Vengatesan Venugopal
Institute for Energy Systems, University of Edinburgh, Scotland, UK

Atul Agarwal
CERES Global, Asia offices, Bangalore, India

ABSTRACT

Waves are characterised by high energy density content, able to provide significant renewable energy contribution. Identification of wave energy potential locations around Europe can provide significant benefits for application of wave energy converters (WECs). With the highest resources placed in the open Atlantic coastlines of Europe, regions such as the Mediterranean are often overlooked.

In this study, with the use of a third generation spectral numerical model, we asses the available, annual and seasonal resource, and investigate the variations that occur within the region. The model utilised, is a nearshore coastal model, which in contrast to oceanic models, has enhanced physical options activated for nearshore and coastal environments, calibrated to the region.

A nested approach was employed, this allowed for in-depth exploration of the resource at coastal environments and more efficient allocation of computational resources. With the nested meshes having a spatial resolution of 0.025°, allowing higher degree of coastal representation.

This resulted to a validated detail resource assessment spanning from 1980-2014 initially for the whole of the Mediterranean, and subsequently for six countries separately. Amongst them, areas for which the wave energy resource was never quantified. The model is calibrated and validated, with results similar to recent studies (Liberti et al. 2013, Zacharioudaki et al. 2015), while it offers improved results by a prior study conducted by (Lavidas et al. 2014). The final resource assessment results are not restricted only to specific areas or short duration, adding to the wave studies performed so far in the region, with a dedicated long-term wave power analysis.

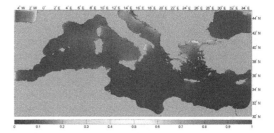

Figure 1. Covariance of Wave Power kW/m.

While the energy levels are not as high as the open oceanic waters, the standard deviation and the covariance of the region presents reduced values. With the energy content having less volatility per year, see Figure 1. Seasonally, winter and autumn periods (as expected), present the most interesting production months. The combination of the high spatial resolution, and temporal duration allows us to investigate several locations and assess their energy suitability.

REFERENCES

Lavidas, G., V. Venugopal, & D. Friedrich (2014). Investigating the opportunities for wave energy in the Aegean Sea. In *7th Int. Sci. Conf. Energy Clim. Chang. 8–10 Oct. 2014 Athens*, Athens. PROMITHEAS The Energy and Climate Change Policy Network.

Liberti, L., A. Carillo, & G. Sannino (2013, feb). Wave energy resource assessment in the Mediterranean, the Italian perspective. *Renew. Energy 50*, 938–949.

Zacharioudaki, A., G. Korres, & L. Perivoliotis (2015). Wave climate of the Hellenic Seas obtained from a wave hindcast for the period 1960–2001. *Ocean Dyn.* 65(6), 795–816.

Numerical simulation and validation of CECO wave energy converter

P. Rosa-Santos, M. López-Gallego & F. Taveira-Pinto
Interdisciplinary Centre of Marine and Environmental Research (CIIMAR) and Faculty of Engineering of the University of Porto, Department of Civil Engineering, Porto, Portugal

D. Perdigão
Faculty of Engineering of the University of Porto, Porto, Portugal

J. Pinho-Ribeiro
PT Portugal, Porto, Portugal

ABSTRACT

CECO is a new wave energy converter (WEC) designed to convert simultaneously the kinetic and the potential energy of ocean waves into electricity. Its working principle is based on the oblique motion of two lateral floating modules in relation to a fixed element. The proof of concept of this device was successfully carried out at the Hydraulics Laboratory of the Faculty of Engineering of the University of Porto, Portugal (Marinheiro et al., 2015; Rosa-Santos et al., 2015), using a physical model built on a geometric scale of 1/20.

The experimental results obtained are used in the paper to calibrate and validate a hydrodynamic model of CECO created with Ansys® Aqwa™, which is a code based on the Boundary Element Method (Ansys, 2013). A mesh with 2349 panels and 304 Morison elements was used (Figure 1) and two different WEC configurations were tested: with and without the power take-off (PTO) system incorporated.

The numerical results showed a good agreement with the physical model ones and allowed to obtain

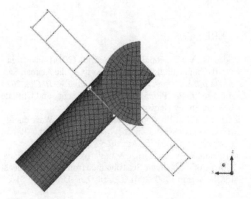

Figure 1. Lateral view of the mesh used for the numerical modeling of CECO.

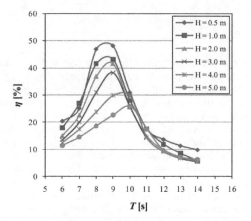

Figure 2. Percentage of the incident wave power captured by CECO with PTO system (values in prototype).

a better insight into the performance of CECO. The efficiency of the WEC capturing the wave energy was analyzed and the results were very promising: values higher than 50% were obtained for some test conditions, Figure 2. Nonetheless, to maximize the energy production, the power delivered to the PTO system should be maximized. The numerical modelling approach followed was found to be adequate, therefore the calibrated CECO model is a solid basis for future numerical optimization of this WEC.

REFERENCES

ANSYS, 2013. Aqwa Theory Manual. ANSYS, Inc., USA, Release 15.0, November 2013.

Marinheiro, J., Rosa-Santos, P., Taveira-Pinto, F. & Ribeiro J., 2015. Feasibility study of the CECO wave energy converter. *Maritime Technology and Engineering*, 1259–1267, doi: 10.1201/b17494-170.

Rosa-Santos, P., Taveira-Pinto, F., Teixeira, L. & Ribeiro, J.P., 2015. CECO wave energy converter: Experimental proof of concept. *J. Renewable Sustainable Energy* 7: 061704.

Pumped hydroelectric energy storage: A comparison of turbomachinery configurations

A. Morabito, J. Steimes & P. Hendrick
Aero-Thermo-Mechanics Departement, Université Libre de Bruxelles, Brussels, Belgium

ABSTRACT

In the past decade, the concerns about world environmental issues produced by fossil fuel exploitation have increased. The use of renewable energy sources aims to reduce the dependence on thermal power plants, that nowadays satisfy most of the global energy demand. Among all the technologies usable for this goal, hydro-power is by far the most used in electric energy production. In such context, the Pumped Hydroelectric Energy Storage (PHES) finds its role as the most mature technology regarding energy storage. PHES systems obviously have many similarities to conventional hydro-power plants but differ by the fact that the flow is bidirectional. A PHES unit exploits the potential energy stored in the upper reservoir as a conventional hydro-power plant, but it converts electric energy from the grid to refill this reservoir, by pumping back water from the lower reservoir when it is economically profitable. The system working conditions are then designed according to the electricity trading and regulations services needed by the local energy market.

In this paper three solutions for PHES are discussed: traditional ternary applications that include separate hydraulic machines, Reversible Pump Turbines (RPTs) and Pumps as Turbines (PATs). RPTs are hydraulic machines designed to work either as a pump or as a turbine depending on the direction of rotation. They can generally benefit of a power-load control by means variable geometry or variable rotation speed. Moreover RPTs leads to a compact power house, saving equipment and civil costs.

PATs emerge as an attractive alternative. In fact, pumps have large market, cheap and reliable. In this paper different prediction models are listed and a revised estimation of the pump performances in reversed operation is given.

A case study compares different system efficiencies and capacity for different configurations. This PHES consist in a unused mine with a depth of 350 meters and it is equipped with one or more Francis turbine or with PAT unit. The available head heavily varies with

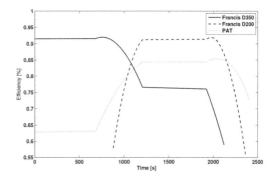

Figure 1. Efficiency trend of the turbines while generating in the case study over time.

interittency due to the peculiar mine configuration. Consequently the hydraulic machines adopted could often work in off-design as showed in Fig. 1. PATs show a value of power generation reduced compared to the other options. Concerning the efficiency, PATs supported by the VFD are able to work more efficiently than turbines in the relevant off-design range and the performance of hydraulic turbines appears to be higher than PATs.

The configurations proposed for the case study had to demonstrate the profitability of this non-conventional PHES plant. Thus the exploitation of mines for PHES is a suitable alternative for those countries do not have significant mountainous ranges.

REFERENCES

Derakhshan, S. & A. Nourbakhsh (2008). Experimental study of characteristic curves of centrifugal pumps working as turbines in different specific speeds. Experimental Thermal and Fluid Science 32(3), 800–807.

Gulliver, J. & R. Arndt (1991). Hydro-power engineering handbook. McGraw-Hill.

Paish, O. (2002). Micro-hydropower: Status and prospects. Proceedings of the Institution of Mechanical Engineers, 216, 31–40.

Cost and revenue breakdown for a pumped hydroelectric energy storage installation in Belgium

J. Steimes, G. Al Zohbi & P. Hendrick
Aero-Thermo-Mechanics Department, Université Libre de Bruxelles, Brussels, Belgium

B. Haut & S. Doucement
Mathematics, Data Processing and Information Technology & EGEU, Laborelec, Linkebeek, Belgium

ABSTRACT

The technical and economic feasibility of Pumped Hydroelectric Energy Storage (PHES) is increasingly under discussion in Belgium and worldwide. The important initial capital cost, a barrier to the use of PHES systems, could be counterbalanced by using pre-existing installations (as mines or quarries). However, it is necessary to have an estimation of the costs and revenues breakdown of these installations in order to optimise the number and the type of components that will be installed.

This paper presents the cost and revenue breakdown for a PHES installation to be installed in Belgium, more specifically in the Walloon region. It presents the division of the plant into sub-parts for which the investment costs are modelled (reservoirs, electrical lines, etc.). An optimisation of the revenues sources available to PHES plants in Belgium is then presented. This helps to optimise the utilisation of the plant. Operation and Maintenance costs are finally then taken into account to build a financial analysis of the investment.

This technico-economical analysis is performed on a quarry and a mine providing 4 MW/20 MWh services to the network. Despite providing the same power and energy levels, the topological configuration is defined differently. The mine presents a high head with a small water volume storage, while the quarry has a low head but a large volume. Results show that there are no large differences in the revenues generated (apart resulting from the efficiency). However, the investment cost change (Fig. 1). Indeed, excavation and spillway costs are higher for a quarry than a mine due to the size of the artificial tanks that needs to be constructed. With the most optimist revenue streams, the net present value of the mine would breakeven in 6 years, while the quarry

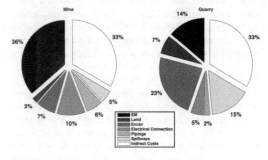

Figure 1. Cost breakdown for a PHES in a mine and a quarry.

would need 13 years. With less optimist hypothesis, only the mine would breakeven after 80 years. Therefore, at this first stage of study with two hypothetical site configurations, it seems more interesting to investigate PHES solution in a mine with an high head and a low volume. This will be discussed in future works as additional costs might arise in mines that are sealed.

REFERENCES

[1] Zhang, Q.F., Smith, B. and Zhan, W., 2012. Small hydropower cost reference model. Technical Report, Oak Ridge National Laboratory, Tennessee, USA.

[2] Agadis, G.A., Luchinskaya, E., Rothschild, R. and Howard, D.C., 2010. The cost of small-scale hydro power production: Impact on the development of existing potential. Renewable Energy, 35, 2632–2638.

[3] Kaldellis, J., Vlachou, D. and Korbakis, G., 2005. Techno-economic evaluation of small hydro power plants in Greece: a complete sensitivity analysis, Energy Policy, 33(15), 1969–1985.

Sustainable Hydraulics in the Era of Global Change – Erpicum et al. (Eds.)
© 2016 Taylor & Francis Group, London, ISBN 978-1-138-02977-4

Numerical investigations of a water vortex hydropower plant implemented as a fish ladder – Part I: The water vortex

N. Lichtenberg
Laboratory of Fluid Dynamics and Technical Flows (LSS), University of Magdeburg "Otto von Guericke", Magdeburg, Germany

O. Cleynen & D. Thévenin
Laboratory of Fluid Dynamics and Technical Fluidics (LSS), University of Magdeburg "Otto von Guericke", Magdeburg, Germany

ABSTRACT

The requirements, which ensure good water quality in natural bodies of water in Europe, also involve restoring the ecosystem and therefore free migration of fish upstream and downstream. As a consequence, measures must be taken and structures must be built which allow for this migration.

This paper provides a very detailed overview of the resulting geometric and hydraulic specifications for such structures. On the basis of the results of a vortex hydropower plant simulation, the plant's suitability as a fish ladder is underlined, as all requirements in terms of flow velocity are met.

The system is home to a complex swirl-prone, unsteady flow with a free surface. In order to ease convergence, the flow is ramped up progressively, raising from 0 kg/s up to 850 kg/s after 10 seconds. In order to optimize computing resources, an automatic mesh refinement at the points of interest such as the free surface and areas high-velocity flow, was performed. The ensuing grid structure can be seen in Figure 2.

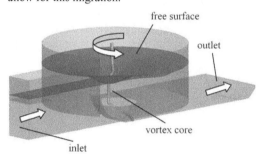

Figure 1. Geometric representation of the water vortex power plant.

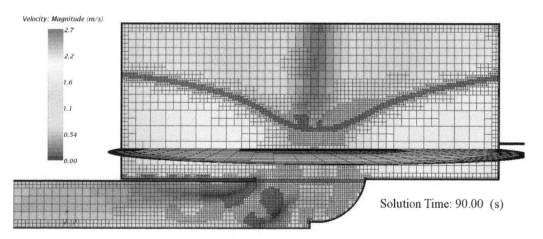

Figure 2. Velocity profile with mesh size distribution after adaptive meshing at regions of high velocity gradient and free surface.

Performance mapping of ducted free-stream hydropower devices

O. Cleynen, S. Hoerner & D. Thévenin
Lehrstuhl für Strömungsmechanik und Strömungstechnik (LSS), Otto-von-Guericke-Universität Magdeburg/ISUT, Magdeburg, Germany

ABSTRACT

A series of numerical experiments is conducted to evaluate the effect of ducting on the hydraulic power production characteristics of a low-impact, low-power, free-surface hydraulic device. A methodology is presented to identify, for a given duct geometry, the relative duct size which maximizes the device power density: this requires systematic coverage of the power map. Results from this calculation-intensive method are compared with a previously-developed theoretical model. The need for a large number of individual experiments or simulations (so that for each size ratio the optimum operating velocity may be captured) is highlighted. Trends identified previously are broadly followed: while increases in power density may be achieved with ducting, no increase in the power coefficient must be expected unless the hydraulic efficiency of the actuating parts directly benefits from the flow condition changes. Nevertheless, the duct drag behavior observed in experiments differs from that of the model, due to the wide range of flow patterns observed across size ratios in these experiments. In future work, a geometrical scaling process may be developed by which the flow regime around the duct is kept unchanged across size ratios, so as to permit accurate analytical modeling. The experimental costs associated with the development of efficient low-impact hydropower devices could then be reduced.

In-situ scale testing of current energy converters in the Sea Scheldt, Flanders, Belgium

T. Goormans & S. Smets
International Marine and Dredging Consultants nv, Antwerp, Belgium

R.J. Rijke
Water2Energy B.V., Heusden, The Netherlands

J. Vanderveken
AquaScrew bvba, Asse, Belgium

J. Ellison
Blue Energy Canada Inc., Richmond, British Columbia, Canada

R. Notelé
Waterwegen en Zeekanaal nv – Sea Scheldt Division, Antwerp, Belgium

ABSTRACT

Waterwegen en Zeekanaal (W&Z) participated in the European project PRO-TIDE, sponsored by the Interreg IVB North-West Europe program. The main goal of this project was "to increase the use of renewable energy by promoting innovative, sustainable and cost effective solutions for tidal energy […], in coastal zones and estuaries". Within PRO-TIDE, W&Z organised, with IMDC providing technical assistance, the testing of different current energy converters.

After a selection procedure, described in Goormans et al. (2015), the bridge in Temse (the first road crossing over the Scheldt counting from downstream), was deemed suitable for testing.

The test set-up comprised of a work pontoon, moored between two piles, piled in the river bed. The turbines were connected to this pontoon in a floating set-up, offering the additional advantage of capturing the highest velocities occurring in the upper layers of the water column. The set-up was completed with two RCM-9 current measurement devices. Finally, the power at the axis of each turbine was measured.

Three different devices were consecutively tested during 4 weeks (Water2Energy, AquaScrew, and Blue Energy Canada). Besides characterizing the performance of the devices, the tests aimed at gaining experience in more practical aspects, such as logistics, installation techniques, and maintenance requirements.

From the data, turbine performance coefficients could be estimated. Performance was rather low, indicating that further technological development is required before it becomes profitable to generate electricity from the relatively low flow velocities in the Sea Scheldt. These low velocities however were found to create a challenging environment to install current energy convertors.

Accumulation of floating debris is a point of attention. Some parts of the turbine assemblies were 'hot spots' for accumulation, causing additional turbulence in the flow.

The conversion from mechanical to electric power requires adequate attention. A suboptimal design of the drive train or power electronics lower the performance, so care should be taken so as to incorporate an efficient and versatile power management system in the design.

There was no systematic monitoring of fish or other animals, so no general conclusions can be drawn. Nevertheless, no signs of collisions or other impacts were observed when visiting the turbines.

Finally, the logistics of even a relatively small prototype test such as performed in this study, should not be underestimated. Constructing a (temporary) set-up in a river requires specialised equipment. Moreover, in tidal rivers, the installation and demobilisation require careful planning, because of the need for low velocities. Slack water conditions are preferable. Therefore, an experienced contractor who knows the local conditions in the river, is vital for a smooth execution.

Despite the low power output, the tests did broaden the knowledge related to harvesting energy from low flow velocities, not only from a design perspective but also from an operational point of view. Moreover, the technology is still in its infancy, meaning that there is room for improvement and optimisation, so it is expected turbine performance factors will increase over time.

REFERENCE

Goormans, T., Smets, S. & Notelé, R. 2015. Is harvesting blue energy feasible in the Sea Scheldt River, Flanders, Belgium? Finding a suitable location for testing different in-river free flow turbines. *Proc. 36th IAHR World Congress*, 28 June–3 July 2015. The Hague, The Netherlands.

Coupling of water expansion and production of energy on public water distribution network

B. Michaux
CILE (Compagnie Intercommunale Liégeoise des Eaux), Liege, Belgium

C. Piro
Cla-Val, Switzerland

ABSTRACT

CILE produces and supplies drinking water to more than half a million people in the Greater Liege area. With a production of 100.000 m^3 of water per day, a distribution network of more than 3.550 km and about 330 civil engineering structures, the Company is one of the major player in the water sector in Wallonia (Belgium).

Such an activity would be not possible without very high professional technical supports such as digitized mapping or management of remote monitoring equipment.

Unfortunately, the smallest structures are not systematically connected to power supplies because of their situations (wood, countryside). But it is very important to send/receive information on the quantity or the pressure in order to guaranty a continuous distribution service.

After discussion with a local distributor of hydraulic equipment in business with a Swiss Company producing among other things control valves, the project consisting in the recovering of the expansion energy in a small water turbine was born.

A flow of 50 l/min with an expansion of 0.6 bar is just enough for the production of energy with a power of 14 W stored in a 12 V & 24 V DC battery. Then, this one supplies a PLC that communicates with the central supervision system.

In the past, the energy was produced with photovoltaic panels in summer and with pre-charged batteries in winter. The new solution allows a reduction of the exploitation costs very quickly because of the high competitive price of the micro-turbine (CLA-VAL e-Power IP), now installed on 3 different networks.

Coastal aspects in the era of global change

Impact study of a new marina on the sediment balance. Numerical simulation

M.C. Khellaf & H.E. Touhami
LEGHYD Laboratory, Faculty of Civil Engineering, University of Science and Technology Houari Boumediene, Algiers, Algeria

ABSTRACT

Numerical modeling of hydrodynamic processes is a powerful tool to study a sandy coast and the consequences of a coastline layout on natural conditions.

This tool allows solving the equations governing water mass movements (fluid mechanics) to know on a given geographic area, fluctuations in time and space of the hydraulic variables for different weather-oceanographic conditions.

In the early 80s the first models of morphological evolution in the combined of waves action and currents (Andersen et al., 1988) emerged. These models use the depth-averaged wave and current equations and can simulate the morphological variation in space and time.

The simulations were performed using the Mike 21 software (Brøker et al., 2007; Zyserman and Johnson, 2002), developed by the Danish Hydraulic Institute (DHI), which has a very high reliability.

This paper presents a preliminary study of the construction of a new marina in the governorate of Boumerdes (Algeria). This study is to determine the potential impact of the planning on the sediment balance of the site.

The review of the initial state where attempts to reproduce numerically the hydro-sedimentary phenomena prevailing on the site. This examination consists of present successively the waves, the induced currents, the theoretical capacity of sediment transport for directions 315° N and 360° N and 45° N.

The model was calibrated by use of field data. The model reproduced hydrodynamics satisfactorily as well as the rate of sediment transport. The bed evolution derived from simulations shows a good agreement with the survey both in the locations and in the intensity of erosive and depositional areas.

After, we will examine the state in the presence of the new marina. For different directions of the waves we will numerically reproduce hydrosedimentary phenomena.

Comparison of the results of these simulations can show the impact of the proposed variant on the hydrosedimentary behavior and inherent risks of silting.

REFERENCES

Andersen, O.H.; Hedegaard, I.B.; Dreigaard, R.; de Girolamo, P., and Madsen, P., 1988. Model for morphological changes under waves and current. *Proc. IAHR Symposium on Mathematical Modeling of Sediment Transport in the Coastal Zone* (Copenhagen. DHI, Horsholm), pp. 310–319.

Brøker, I.; Zyserman, J.; Madsen, E.Ø.; Mangor, K., and Jensen, J., 2007. Morphological modeling: a tool for optimization of coastal structures. *Journal of Coastal Research* 23(5), 1148–1158.

Zyserman, J.A. and Johnson, H.K., 2002. Modelling morphological processes in the vicinity of shore-parallel breakwaters. *Coastal Engineering*, 45, 261–284.

Numerical simulation of scour in front of a breakwater using OpenFoam

N. Karagiannis, T. Karambas & C. Koutitas
Department of Civil Engineering, Aristotle University of Thessaloniki, Thessaloniki, Greece

ABSTRACT

An innovative numerical approach for the sediment transport and scour simulation is applied and presented in this work. It is about a repetitive method which consists of two repeated steps. The first one concerns the implementation of a hydrodynamic model, developed on the OpenFoam platform, while in the second step, a sediment transport model, developed in FORTRAN, is applied using the hydrodynamic results of the first model. The product of the latter model is a new bathymetry, given the hydrodynamic state, which was established after the implementation of the first model. Then, the new bathymetry is applied in the first model and a new hydrodynamic state is arisen, which is used by the second model for a new bathymetry to be created. The above procedure is repeated until convergence, namely the new bathymetry coincides with the previous one and equilibrium of the seabed has been reached.

The aforementioned former numerical model was created with the open source toolbox OpenFoam and the additional toolbox waves2Foam (Jacobsen et al. 2012). The RANS equations have been solved simultaneously with the transport equations of the turbulence model k-ω SST and the Volume of Fluid method's ones.

The second numerical model has been developed in FORTRAN, estimating the sheet flow sediment transport rates with the Camenen and Larson (2007) transport rate formula, as well as the bed load and suspended load over ripples (Karambas, 2012). Suspended sediment transport rate is incorporated by solving the depth-integrated transport equation for suspended sediment (Karambas, 2012). After that, the conservation equation of the sediment mass is applied for several time steps and the scour is computed.

The numerical experiment concerns a case, whose geometry is a wave channel, where a vertical breakwater is lying at 500 m from the left boundary of the channel. A cnoidal wave of H = 1 m height and T = 6 s period is applied.

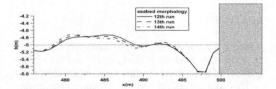

Figure 1. Seabed morphology after the last run.

Fourteen steps were required until convergence and the scour depth takes its final value of about 0.75 m. It was the last 3 steps that seem to yield identical results (Fig. 1) in terms of the seabed morphology and the scour formation.

The hydrodynamic OpenFoam model seems to behave very well with respect to the standing wave formation and evolution, while the second one yields realistic seabed morphology at every step.

In conclusion, the present numerical approach seems to behave very well from qualitative point of view, hence a comparison with experimental results will be the next step in order for this method to be validated.

REFERENCES

Camenen, B. & Larson, M. 2007. A unified sediment transport formulation for coastal inlet application, *Technical report ERDC/CHL CR-07-1, US Army Engineer Research and Development Center*. Vicksburg, MS.

Jacobsen, N.G., Fuhrman, D.R., & Fredsøe, J. 2012. "A Wave Generation Toolbox for the Open-Source CFD Library: OpenFoam". *Int. J. Numerl. Meth. Fluids*, 70(9): 1073–1088.

Karambas, T.V., 2012. Design of detached breakwaters for coastal protection: development and application of an advanced numerical model, *Proceedings of the 33rd International Conference on Coastal Engineering 2012*, 1(33), sediment.115. doi:10.9753/icce.v33.sediment.115

Application of an unstructured grid tidal model in the Belgian Continental Shelf

Biniyam B. Sishah & Margaret Chen
Department of Hydrology and Hydraulic Engineering, Vrije Universiteit Brussel, Brussels, Belgium

Olivier Gourgue
Department of Hydrology and Hydraulic Engineering, Vrije Universiteit Brussel, Brussels, Belgium
Flanders Hydraulics Research, Flemish Government, Antwerp, Belgium

ABSTRACT

Performance analysis for unstructured grid with variable mesh densities covering the entire Northwestern European continental shelf area was made in TELEMAC-2D model v7p0r1, for the calm period (in terms of wind) of July 2013. The aim was to calibrate the model for tides, before modeling the impacts of super storms in the Belgian Continental Shelf (BCS) area.

Accordingly, in setting up of the TELEMAC-2D model, the meteorological forces were not considered and a constant value of the Coriolis coefficient suited for the BCS was used. In terms of boundary conditions, water depths and velocities are imposed at the shelf break using the regional OTIS tidal solution on the European continental shelf, providing amplitudes and phases of 11 tidal harmonic components with a spatial resolution of 2 arc-minutes. The same dataset was used for the initial conditions.

For large computational domains such as the one considered for this model, the use of spherical coordinates and a space varying Coriolis coefficient are important. However, that was not possible with the version of TELEMAC-2D used for this study (v7p0r1), due to a bug, now corrected in the current version (v7p1).

Statistical goodness of fit tests comparing model results and reference dataset (European dataset of OTIS) tidal constituents were made for the calibration process. Constituents considered in the analysis were M_2, S_2, N_2, O_2, Q_1, K_1, MN_4 and M_4. Primarily, model performance in the area of the open boundary was assessed considering that computations actually progress from this part of grid and highly influence results in the entire domain. Analysis results have indicated that tidal constituents had good performance, while a slight reduction in performance were observed for both velocity component constituents. Further works are recommended on this part.

Sensitivity analysis made for model parameters have indicated that bottom friction coefficient was the most sensitive. Practical ranges of the Manning coefficient were tested in model and $0.024 \, s/m^{1/3}$ was found to be the optimum value for the coefficient. For this parameter value, good performance was achieved in deep waters of the BCS, while medium performance is observed near the Scheldt Estuary region. It is expected that inadequate representation of downstream rivers of the Scheldt Estuary in model's grid might have played a role. Overall, model have shown good performance in the BCS.

Time series analysis were also incorporated in this study, tidal level measurements obtained from five observation stations in the BCS area were compared to corresponding model results. Analysis results have shown that the TELEMAC-2D model calibrated here has achieved a performance level higher than that attained by the reference dataset in the BCS location. Hence, the model is now ready to perform simulations considering the effects of wind, air pressure and tidal waves to assess future impacts of super storms in the study area.

Online coupling of SWAN and SWASH for nearshore applications

M. Ventroni & A. Balzano
Department of Civil, Environmental and Architecture Engineering, Faculty of Cagliari, Cagliari, Italy

M. Zijlema
Environmental Fluid Mechanics Section, Faculty of Civil Engineering and Geosciences, Delft University of Technology, The Netherlands

1 INTRODUCTION

An online, one-way coupling procedure between the SWAN phase-averaged, spectral wave model (Booij et al. 1999,) and the SWASH time domain, multi-layered non-hydrostatic model (Zijlema et al. 2011) has been developed, with the aim of modeling, seamlessly, efficiently and accurately, the wave evolution from generation to runup and land inundation.

SWASH is forced by directional wave random fields generated with a wave-maker algorithm (Zijlema et al. 2011) from the action density spectra computed by SWAN at its open boundary every times step, which is typically much larger than SWASH time step. This assures continuity of information, while preventing spurious reflection of outgoing waves. The wave-induced setup calculated by SWAN is also passed to SWASH.

The coupling efficiency has been evaluated through comparison with laboratory data (Mase 1989), see Table 1, with special attention paid to the optimal location of the inner boundary (McCabe et al. 2011) in terms of the nonlinear parameter H_{m0}/d.

Runup statistics are reasonably well predicted as summarized in Figure 1, with optimal coupling point obtained for $H_{m0}/d \approx 0.3 - 0.55$, noticeably smaller than $H_{m0}/d \approx 0.65$ found by McCabe et al. (2011), probably due to SWASH dispersive properties.

Table 1. Parameters of incident waves at the wavemaker for the Mase (1989) tests.

Case	H_{wm} (cm)	T_p (s)	L_{wm} (cm)	ξ_{wm} (–)	Breaker type
TEST A	4.95	2.50	500	0.702	Plunging
TEST B	6.18	2.00	388	0.502	Spilling
TEST C	7.37	1.67	312	0.384	Spilling
TEST D	9.14	1.25	212	0.258	Spilling

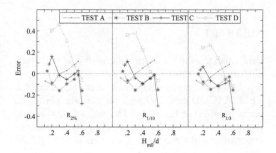

Figure 1. Errors in runup statistics for the coupled SWAN+SWASH model, compared with Mase's (1989) laboratory data, as a function of H_{m0}/d at coupling location. Errors are mean values of 5 randomly phased level signals obtained from the same spectra.

REFERENCES

Booij, N., Ris, R.C. & Holthuijsen, L.H. 1999. A third-generation wave model for coastal regions. 1. model description and validation. *Journal of Geophysical Research* 104(C4): 7649–7666.

Mase, H. 1989. Random wave runup height on gentle slope. *Journal of Waterway, Port, Coastal and Ocean Engineering* 115(5): 649–661.

McCabe, M., Stansby, P.K. & Apsley, D.D. 2011. Coupled wave action and shallow-water modeling for random wave runup on a slope. *Journal of Hydraulic Research* 49(4): 515–522.

Zijlema, M., Stelling, G.S. & Smit, P. 2011. SWASH: an operational public domain code for simulating wave fields and rapidly varied flows in coastal waters. *Coastal Engineering* 58: 992–1012.

Spectral parameters modification in front of a seawall with wave return

Th. Giantsi, G. Papacharalampous, N. Martzikos & C.I. Moutzouris
Laboratory of Harbour Works, School of Civil Engineering, National Technical University of Athens, Greece

ABSTRACT

Seawalls with wave return are vertical structures constructed to protect the inland from the wave action and from the wave overtopping. Simultaneously, seawalls affect incident waves by reflecting them.

To investigate the performance of a composite seawall with wave return, a 3 m-long physical model was constructed in a wave basin at the Laboratory of Harbour Works, National Technical University of Athens. The physical model of the seawall was constructed on a horizontal bottom, reproducing the foundation of the structure only till a constant depth. The waves were measured in front of it, at four locations. A 5th wave gauge was located near the wave generator to control the incident wave.

Two different sections of the composite seawall were analyzed, under a combination of water levels and wave conditions. A typical cross section of the experimental layout is presented in Figure 1.

Spectral analysis of time series of sea surface elevations by spectral analysis provides as main results the moments (zero to nth order) which describe the spectrum. Using the moments, dimensional and dimensionless parameters were calculated to provide further information on the spectrum. Common dimensional spectral parameters are the significant wave height H_s, the mean period T_m, and the wavelength L_m. Where:

$$\varepsilon = \sqrt{1 - \frac{m_2^2}{m_0 m_4}}, \quad \nu = \sqrt{\frac{m_2 m_0}{m_1^2} - 1}$$

To better describe the spectral shape, three dimensionless shape parameters were used: the broadness parameter ε, the narrowness parameter ν and the peakedness parameter Q_p. They are defined as follows:

$$H_s = 4\sqrt{m_0}, \quad T_m = \sqrt{\frac{m_0}{m_2}} \text{ and } L_m = g\frac{T_m^2}{2\pi} \text{ and}$$

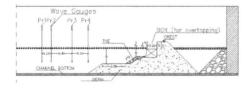

Figure 1. Layout of the model.

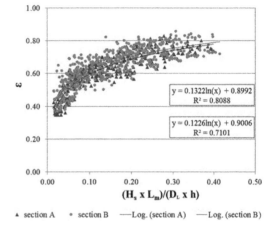

Figure 2. Parameter ε versus d*.

$$Q_p = \frac{2}{m_0^2} \int_0^\infty f[S(f)]^2 \, df$$

These parameters are correlated to the Ursell number (U_r) and to the dimensionless parameter d*, where:

$$U_r = H_s \times L_m^2/h^3 \quad \text{and} \quad d^* = \frac{H_s \times L_m}{D_L \times h}$$

The calculated parameter ε is presented in Figure 2.

The main findings are the following. (a) Parameters ε and ν depend on Ursell number and the parameter d*. The parameter d* is proportional to the wave height, to the wave length and inversely proportional to the distance from the structure and to the water depth. (b) The parameter Q_p depends proportionally on the wave length and on the distance from the structure, is inversely proportional to the water depth and is independent from the wave height. (c) Deformation of the spectrum shape was observed.

Further research is needed in order to incorporate the influence of the structure shape (including the freeboard) on the spectrum shape and on the spectral parameters.

Renewal time scales in tidal basins: Climbing the Tower of Babel

Daniele P. Viero & Andrea Defina
Department of Civil, Architectural and Environmental Engineering (ICEA), University of Padova, Padova, Italy

Abstract

Transport time scales are key parameters to assess water renewal of estuaries and tidal basins and they have long been used to this purpose. They have become more and more popular with the growing use of numerical models, since renewal time scales are rational tools to "condense" huge amount of numerical data into intelligible, still quantitative, information (Deleersnijder and Delhez 2007). However, growing up is never straight forward: in a modern-day Tower of Babel, an ever-increasing confusion has involved terminology, use and estimation of time scales, despite valuable efforts towards a sounder theoretical framework (from Bolin and Rodhe 1973 to Delhez et al. 2014). This paper is a modest contribution to this sort of "climbing" toward a well-established definition, and use, of renewal time scales.

We focus on the Venice Lagoon (Italy), a ≈ 550 km^2 shallow, micro-tidal basin in the northern Adriatic Sea (Fig. 1). We thus focus on tidally flushed, semi-enclosed basins with negligible freshwater inflows, in which water renewal is strongly controlled by diffusion. The inlets act alternatively as source and sink, and a significant fraction of effluent water can return to the basin each flood tide. Hence, to correctly assess water renewal of such basins, a region larger than the basin itself has to be considered (Fig. 1).

To account for the effects of return flow, the concept of exposure time was introduced as an alternative to the residence time (Monsen et al. 2002). However, deep analysis of exposure time revealed unexpected complications that, in part, are still open questions (Delhez 2013, Viero and Defina 2016).

In the present contribution, both water age and local flushing time account for the return flow, as for exposure time. These time scales are defined and compared to each other. Approximations to the real time scales are introduced, as an attempt to reduce the dependence of the mean time scales on the location of boundary conditions. The results provided by a two-dimensional, mixed Eulerian-Lagrangian, semi-implicit finite element model, help in comparing the different time scales and in assessing the accuracy of the proposed approximations. We show that the local flushing time well approximates, quantitatively, both water age and exposure time in steady periodic hydrodynamic condition and when advection is negligible. Conversely, wind-driven advection makes the exposure time to be significantly different from water age and local flushing time.

Figure 1. The Venice Lagoon, in the north-eastern Italy.

REFERENCES

Bolin, B. & H. Rodhe (1973). A note on the concepts of age distribution and transit time in natural reservoirs. *Tellus* 25, 58–62.
Deleersnijder, E. & E. Delhez (2007). Timescale- and tracer-based methods for understanding the results of complex marine models. *Estuar. Coast. Shelf. S.* 74, v–vii.
Delhez, E. (2013). On the concept of exposure time. *Cont. Shelf Res.* 71, 27–36.
Delhez, E., B. de Brye, A. de Brauwere, & E. Deleersnijder (2014). Residence time vs influence time. *J. Marine Syst.* 132, 185–195.
Monsen, N., J. Cloern, L. Lucas, & S. Monismith (2002). A comment on the use of flushing time, residence time, and age as transport time scales. *Limnol. Oceanogr.* 47(5), 1545–1553.
Viero, D. & A. Defina (2016). Water age, exposure time, and local flushing time in semi-enclosed, tidal basins with negligible freshwater inflow. *J. Marine Syst.* 156, 16–29.

Morphodynamic processes at Mariakerke: Numerical model implementation

R. Silva
Flanders Hydraulics Research, Belgium
Antea Group, Belgium

R. De Sutter
Antea Group, Belgium

R. Delgado
Flanders Hydraulics Research, Belgium

ABSTRACT

Within the framework of the project "Alternative maintenance measures for beach nourishment: monitoring of a pilot site", the contribution of underwater nourishment to reduce maintenance, and thus ex-tending the lifetime of beach nourishment along the Belgium coast, without compromising coastal safety, is to be assessed. This is based on an integrated approach, including field measurement campaigns, numerical modelling and the definition and evaluation of a number of morphological indicators, for Mariakerke coastal area in Ostend, Belgium. Within the first steps of the assessment was the implementation of a numerical model for the simulation of the hydrodynamic and morphodynamic processes, occurring during the short term (beach response to storms and nourishments).

XBeach (Roelvink et al., 2010) was set-up and calibrated for high energetic hydrodynamic, sedimentary and morphological conditions, observed during Sinterklaas storm in December 2013 at section 104 of the Belgium coast (Mariakerke).

XBeach was firstly set-up in 1D hydrostatic mode and its performance investigated, using as indicators the profile shape and the erosion volume above storm surge level. A sensitivity analysis to 9 input parameters with a large influence on the calculations and a rank for their influence is presented. The knowledge of the model response to the varying parameters allowed for its calibration. A simulated profile very close to the measured post storm profile and an identical erosion volume above storm surge level were achieved (Figure 1).

Then, the model was set-up in Q2D (alongshore uniform) hydrostatic mode for the same conditions. The comparison of 1D and Q2D simulations showed that there is a tendency of less dune erosion and less wet beach accumulation in Q2D. Applying the settings defined through 1D calibration, resulted in a better model performance than default settings and a better estimate of the erosion volume above storm surge level. The effect of the wave and computational

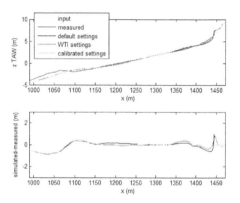

Figure 1. Measured profiles before and after Sinterklaas storm at section 104 of the Belgian coast. Simulated profiles using default, WTI and calibrated settings and differences from measurements.

grids resolutions was inspected. As expected, simulation time decreased with grids resolution decrease. In general, for the tested conditions, decreasing wave grid resolution for short waves and rollers at the offshore boundary, results in enhanced erosion. Also, the cross-shore resolution of the spatial grid has a major influence in the results.

Other periods interesting for model implementation at the study site would be: March–April 2014 and October–November 2014, when a storm with exceptional water levels in spring tide occurred. However, none of the periods is suitable for model validation, since the first corresponds to moderate conditions, with no significant impact on beach morphology, and there was no beach survey shortly before the second.

REFERENCE

Roelvink, D., Reniers, A., van Dongeren, A., van, J., de Vries, T., Lescinski, J., McCall, R., 2010. XBeach Model Description and Manual. Unesco-IHE Institute for Water Education, Deltares and Delft.

Design features of the upcoming coastal and ocean basin in ostend, Belgium

A. Kortenhaus, P. Troch, N. Silin, V. Nelko, P. Devriese, V. Stratigaki & J. De Maeyer
Department of Civil Engineering, Ghent University, Zwijnaarde (Ghent), Belgium

J. Monbaliu, E. Toorman & P. Rauwoens
Department of Civil Engineering, KU Leuven, Leuven, Belgium

D. Vanneste, T. Suzuki, T. Van Oyen & T. Verwaest
Flanders Hydraulics Research, Antwerp, Belgium

ABSTRACT

The new Coastal and Ocean Basin (COB) at the Greenbridge Science Park in Ostend, Belgium is under design. The laboratory will provide a versatile facility that will make a wide range of testing possible, including the ability to generate waves in combination with currents and wind at a large range of model scales. This facility is part of the Gen4Wave project on offshore renewable energy and coastal engineering in Flanders, Belgium. The COB is funded by the Hercules foundation, the Agency for Innovation by Science and Technology (IWT), Ministry of Public Works and Mobility, Ghent University (UGent) and University of Leuven (KU Leuven). The basin will be part of a larger building complex that will also host a towing tank from the Maritime Access Division of the Ministry of Mobility and Public Works. This new infrastructure will offer the opportunity to companies and government agencies to develop innovative designs thereby strengthening the position of Flanders in coastal engineering and offshore renewable energy.

The project aligns perfectly with the action plan of the province of West-Flanders, Factories of the Future 'Blue Energy' in supporting developments in the blue energy field. The Greenbridge Science Park is a breeding ground for Blue Growth, and is located in the Port of Ostend where it fits within the focus of the port on renewable energy. The operational management of the infrastructure will be done by the partnership UGent, KU Leuven and Flanders Hydraulics Research.

The COB will be 30 m long, 30 m wide and have a variable water depth of up to 1.4 m. A central pit will allow experiments e.g. with mooring lines, at a depth in excess of 4 m. The basin will contain 2.3 million litres of fresh water and will house an underground water reservoir that allows for quick filling and emptying of the wave basin. The water transfer system will comprise a pumping system with control logic which will achieve the desired water level in the basin.

The basin will be equipped with a unique combination of wave and current characteristics, in order to realize the challenging research roadmap sketched by UGent and KU Leuven. To achieve this purpose, the basin will have state-of-the-art generating and absorbing wave makers and submerged bidirectional propellers that drive the recirculation current across the underground current tank. The wave paddles are planned to be 0.5 m wide and can produce regular waves with a wave height of 0.5 m for a wave period of 3 s, and random waves with a significant wave height of 0.3 m. It will be possible to generate wave-current interactions in the same, opposite and oblique directions and to reach current velocities of 0.4 m/s. The design of the current generation system has been achieved using both numerical and experimental models. Flow velocity fluctuations are expected to be smaller than 10% RMS. For wind generation, a portable device capable of generating speeds up to 15 m/s is foreseen. The basin will be equipped with state-of-the-art instruments for wave, current, wind, kinematic and structural measurements.

The facility is designed to have minimal operating costs while producing high quality experimental conditions. This includes being able to construct coastal landscapes efficiently. For example, it will be possible to access the wave basin with an electric forklift or wheel loader making the model construction process as simple as practically possible. A 10 t overhead crane will enable easy installation of structures and devices.

The COB will allow users to conduct tests for national and international coastal and offshore engineering projects. The basin is expected to be operational in 2017. This paper presents an overview of the basin's capabilities, the ongoing works, and presents and discusses some results of the design work.

Idealized model study on tidal wave propagation in prismatic and converging basins with tidal flats

Thomas Boelens & Tom De Mulder
Department of Civil Engineering, Hydraulics Laboratory,
Faculty of Engineering and Architecture, Ghent University, Belgium

Henk Schuttelaars
Delft Institute of Applied Mathematics, Delft University of Technology, The Netherlands

George Schramkowski
Flanders Hydraulics Research, Belgium

ABSTRACT

The tidal flow in an estuary, and consequently the morphodynamics and ecology, is highly dependent on the basin geometry (Ridderinkhof et al. 2014). The characteristics of tidal wave propagation in prismatic basins are well known, and also the effect of converging basins has been studied to some depth. The influence of tidal flats on tidal wave propagation has been studied using 1D models, while in this paper, both a 1D and a 2D model will be applied. The latter approach allows to account for the water motion above the tidal flats, in contrast to a 1D model where the tidal flats are modelled as storage areas.

To gain some fundamental insight, an idealized model (which is fast, compared to complex process-based models and easy to analyse) has been used, that describes the water motion in a semi-enclosed (converging) basin by means of the shallow water equations, forced by prescribed free surface elevations at the entrance ($x = 0$). The main focus of this paper is to study the influence of tidal flat geometry on the spatial structure of different tidal harmonics and of tidal asymmetry between ebb and flood periods. This work represents a first, validation step in the development of a 2D idealized model for the identification of morphodynamic equilibria. The Finite Element Method (FEM), was used to spatially discretize the governing equations, in which the physical variables are expanded in their tidal constituents. The rectangular and converging tidal basins, that are considered here, can thus be easily extended to more general geometries.

After a favorable comparison with other models in literature, the tidal hydrodynamics was studied for

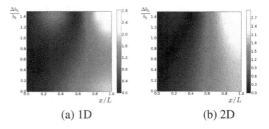

(a) 1D (b) 2D

Figure 1. Amplitude ratio A_{M_4}/A_{M_2} [−] of the free surface elevation as a function of x/L, the normalised distance along the channel, wtih entrance at $x/L = 0$, for different normalised widths of tidal flats $\Delta b_0/b_0$ (b_0 is the width of the main channel at the entrance, Δb_0 is the width of the tidal flats at the entrance).

different values of bottom roughness and for different widths of the tidal flats. As an example, the free surface elevation amplitude of the internally generated overtide M_4 (quarter-diurnal), relative to that of the M_2 (semi-diurnal) component, prescribed at the entrance, is given in Fig. 1, for a converging channel. Though discrepancies exist between the 1D and 2D results, both models show that the amplitude ratio increases towards the closed end of the basin ($x = L$).

REFERENCE

Ridderinkhof, W., H. de Swart, M. van der Vegt, N. Alebregtse, & P. Hoekstra (2014). Geometry of tidal inlet systems: A key factor for the net sediment transport in tidal inlets. *Journal of Geophysical Research: Oceans 119*(10), 6988–7006.

Safety standards for the coastal dunes in The Netherlands

R.P. Nicolai, M. Kok, A.W. Minns & S. van Vuren
HKV Consultants, Lelystad, The Netherlands

C.J.J. Eijgenraam
CPB Netherlands Bureau for Economic Policy Analysis, The Hague, The Netherlands

ABSTRACT

In 2014, the Dutch Delta Programme introduced new safety standards for the Dutch primary flood defence system that were derived using a risk-based approach. The Dutch primary flood defence system consists of a network of dykes (US: levees) and coastal dunes. The new safety standards take into account both the probability of inundation through flooding and the consequences of such a flood. The Dutch Delta Programme has derived the safety standards based on the risks for individual and group mortality (solidarity principle) and on a cost-benefit analysis (i.e. economic optimum; see Figure 1).

A full cost-benefit analysis (CBA) is a data-intensive and a labour-intensive method, and therefore a simple, pragmatic approximation formula, referred to as the direct assessment (DA) approach, has been applied. The DA approach yields a good estimate for economically optimal safety standards for dykes in the Netherlands. The DA approach has subsequently been applied to the coastal dunes in the Netherlands. However, the correctness and applicability of this approximation to these locations has not been based upon hard empirical or analytical evidence. This paper shows that the DA approach is indeed valid for application to coastal dunes in the Netherlands but not as a direct consequence of its applicability to dykes but rather due to the specific nature of sand nourishments, which are used to reinforce the coastal dunes.

The governing assumptions of the DA approach applied to dyke reinforcement require:

1. the water defence under consideration consists of a single section,
2. the optimal size of a reinforcement action is an action that reduces the flood probability of the section by about a factor 10,
3. the average reinforcement costs are almost constant near the optimum, and
4. the failure probability of dykes is log-linear in the dyke heightening.

We demonstrate that after adapting the above assumptions to the case of coastal dunes the DA

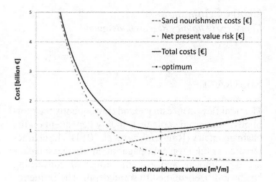

Figure 1. Illustration of the cost-benefit analysis for flood protection of coastal dunes. The x-axis displays the sand nourishment volume (m3/m). The y-axis shows the costs (euros). Red line: investment costs. Blue line: NPV of expected flood damage (economic risk). Black line: total costs.

approach is indeed applicable to coastal dunes, because:

1. the water defence under consideration consists of a uniform length of coast,
2. the average sand nourishment costs are approximately constant,
3. the failure probability of coastal dunes is log-linear in sand nourishment actions.

It also follows that the size of the optimal sand nourishment action is not relevant for deriving the economically optimal flood safety standard. The DA approach does not give any insights into the optimal sand nourishment volume. The optimal sand nourishment volume is most likely small, because the fixed costs are low compared to the variable costs, which depend only on the required sand nourishment volume per meter of coastline. Hence, frequent sand nourishment actions seem to be economically optimal. This corresponds with the present coastal maintenance practice of small-scale sand nourishment actions for the Dutch coast.

Impact of sea level rise on low lying areas of coastal zone: The case of Batu Pahat

A.B. Yannie, A.H. Radzi & A. Dunstan
Coastal Management & Oceanography Research Centre, National Hydraulic Research Institute of Malaysia (NAHRIM), Ministry of Natural Resources & Environment (NRE), Selangor, Malaysia

W.H.M. Wan Mohtar
Department of Civil And Structural Engineering, Faculty Of Engineering, University Kebangsaan Malaysia (UKM), Selangor, Malaysia

ABSTRACT

This study investigates the impact of sea level rise at the coastal zone area of Batu Pahat, Johor which identified as low lying area, located at the south west of Peninsular Malaysia. The coastline of Batu Pahat has experiencing severe erosion and significant land loss due to being inundated by sea water. The inundation profile was predicted using Numerical modelling developed by Danish Hydraulic Institute (DHI) familiarly known as MIKE 21 Flexible Mesh – HD. The model predict the inundation profile for the year of 2020 and 2040 along the Batu Pahat coastline, where the hydrodynamic condition in 2013 was referred as baseline year. The model boundary was design meeting the interest or critical area of the coastal segment of Batu Pahat with dimensions of $42.7 \times 126\,km^2$. Tidal, current and wave data were calibrated and the error ranges are within the acceptable limit. From the model simulation, current speed has increase from 0.2–0.39 m/s (baseline) to 0.36–0.9 m/s in year 2020 and has reached up to 1.8 m/s by the year 2040. On the other hand, inundation map produced showing 1.08% of the developed areas including industrial zone will be affected by the year 2020. The percentage for population area, road networks and mangroves forest is approximately 2.04%, 18.18% and 29.08% respectively for the year 2020. Hitting the year 2040 however, the numbers are expected to increase up to 2.47% for population, 21.21% for road network and about 33.3% for mangrove forest areas.

REFERENCES

Anthoff, D., Nicholls, R.J., Tol, R.S.J. and Vafeidis, A.T. (2006). Working Paper 96. Tyndall Centre for Climate Change Research. *Global and regional exposure to large rises in sea-level: a sensitivity analysis*. Norwich: 2.

Beckman, R. C., 2001. *Using Article 43 of UNCLOS to Improve Navigational Safety and Prevent Pollution in International Straits*. Transboundary Environmental ISSUES:1) Unwanted Stowaway, 2) Ultra Harzardous Cargo, 3) Oil Spills, 4) Navigational Mishaps. Tropical Coasts Buletin, Vol. 8, No. 1. Executive Chairman, Society of International Law, Singapore.

Chia Lin Sien, 1998. *The Importance of The Straits Of Malacca And Singapore*. Singapore Journal Of International & Comparative Law (1998) 2. Pp 301–322.

Cruz R V, Harasawa H, Lal M, Wu S, Anokhin Y, Punsalmaa B, Honda Y, Jafari M, Li C and Huu Ninh N. (2007). Contribution of Working Group II to the Fourth Assessment Report of the Intergovernmental Panel on Climate Change, Parry M L, Canziani O F, Palutikof J P, van der Linden P J and Hanson C E (eds). *Asia- Climate Change 2007: Impacts, Adaptation and Vulnerability*. Cambridge University Press, UK: 469–506.

EPU (1985). National Coastal Erosion Study (NCES). *Final Report, Vol. 1, August 1985*. Kuala Lumpur: Economic Planning Unit (EPU)

IPCC, 1996a. In: Houghton, J.T., Meira Hilho, L.G., Callander, B.A., Harris, N., Kattenberg, A., Maskell, K. (Eds.). 1996a. *Climate Change 1995: The Science of Climate Change*. Cambridge Univ. Press, UK: 572.

IPCC (1996b). In: Watson, R.T., Zinyowera, M.C., Moss, R.H. (Eds.). *Climate Change 1995: Impacts, Adaptations and Mitigation of Climate Change: Scientific –Technical Analyses*. Cambridge Univ. Press, Cambridge: 878.

Mohd Fauzi Mohamad, Lee Hin Lee, and Mohd Kamarul Huda Samion (2014). International Journal of Environmental Science and Development. *Coastal Vulnerability Assessment towards Sustainable Management of Peninsular Malaysia Coastline, December 2014* 5(6): 6.

NAHRIM (2010). *The study of the impact of climate change on sea level rise in Malaysia (Final Report)*. Seri Kembangan, Selangor: National Hydraulic Research Institute Malaysia (NAHRIM).

NPR (2007). Environment. *Study: 634 Million People at Risk from Rising Seas.*, March 28, 2007: Nell Greenfieldboyce.

Rowley, R.J., Kostlenick, J.C., Braaten, D., Li, X. and Meisel, J. (2007). Risk of rising sea level to population and land area. EOS Transactions, 88, 105–107.

Small, C., Gornitz, V., Cohen, J.E. (2000). Coastal hazards and the global distribution of human population: Environ. Geosci. 7, 3–12.

UCS (2013). *Causes of Sea Level Rise. Fact Sheet. April 2013*. Union of Concerned Scientists (UCS), Cambridge, MA: www.ucsusa.org/sea levelrisescience.

Luqman Haziq (2010). *Kebun kelapa 'ditelan' laut*. Johor: Utusan Online.

A database of validation cases for tsunami numerical modelling

D. Violeau, R. Ata, M. Benoit[1] & A. Joly
EDF and Saint-Venant Hydraulics Laboratory, Chatou, France
[1]*Now at IRPHE and Ecole Centrale Marseille, Marseille, France*

S. Abadie, L. Clous, M. Martin Medina & D. Morichon
Université de Pau et des Pays de l'Adour, Anglet, France

J. Chicheportiche & M. Le Gal
Ecole des Ponts ParisTech, Marne-la-Vallée, France

A. Gailler, H. Hébert & D. Imbert
CEA, DIF, DAM, Arpajon, France

M. Kazolea, M. Ricchiuto
Team CARDAMOM, Inria Bordeaux sud-Ouest, Bordeaux, France

S. Le Roy, R. Pedreros & M. Rousseau
BRGM, Orléans, France

K. Pons, R. Marcer & C. Journeau
Principia, La Ciotat, France

R. Silva Jacinto
Ifremer, Géosciences Marines, Centre de Brest, France

ABSTRACT

This work has been performed by a French national consortium within the framework of the national project Tandem (2014–2017), with aim to improve knowledge about tsunami risk on the French coasts. The first of the four work-packages of the Tandem project (WP1 – Qualification and validation of numerical codes) was the opportunity to build a database of benchmark cases to assess the capabilities of various (industrial or academic) numerical codes. 18 codes were used, solving various set of equations with different numerical methods. 14 test cases were defined from the existing literature with validation data from reference simulations, theoretical solutions or lab experiments. They cover the main stages of tsunami life: 1) generation (from seism or landslide), 2) propagation, 3) runup and submersion, and 4) impact. For each case several of the numerical codes were compared in order to identify the forces and weaknesses of the models, to quantify the errors that these models may induce, to compare the various modelling methods, and to provide users with recommendations for practical studies. During the conference, 4 to 5 representative cases will be selected and carefully presented with an analysis of the results. As an example, case P02 is briefly depicted on Figure 1.

Figure 1. Example of tsunami impact test case, from the experiment by Park *et al.* (2013). Top: a VOF method; bottom: an SPH approach.

REFERENCE

Park H., Cox D.T., Lynett P.J., Wiebe D.M., Shin S. 2013. Tsunami inundation modeling in constructed environments: A physical and numerical comparison of free-surface elevation, velocity, and momentum flux. *Coast. Engng.* **79**: 9–21.

Wave force calculations due to wave run-up on buildings: A comparison of formulas applied in a real case

F.G. Brehin, N. Zimmermann, V. Gruwez & A. Bolle
IMDC, Antwerp, Belgium

ABSTRACT

As part of a flood risk study in Northwest France, wave force calculations on schematized buildings located at a varying distance in the wave run-up zone were performed for a total of 33 measured profiles. The choice of the formulas from the literature review included the following: Goda-Takahashi, Camfield, Pedersen (USACE, 2006), Chen (Chen, 2012 and 2015), and a theoretical formula based on EurOtop (Pullen *et al.*, 2007). Each formula is characterised by a particular application range, asymptotic behavior and specific input parameters. All the above formulas enable the calculation of a horizontal force (F_h) except for Eurotop, for which the force was derived from the overtopping flow depths and velocities. Input parameters for the wave force calculations included both hydrodynamic and geometric data. The hydrodynamic data consisted of wave and water level outputs from previous numerical model studies (Alp'Géorisques and IMDC, 2014) for the reference storm event Xynthia (2010). For the geometry, the profiles were schematised manually to determine the slope (α), freeboard (R_c) and crest width (B). An example of the results (cf. Figure 1) shows the beach profile (top) along with the corresponding wave force distribution (bottom), the applicability range (thick lines) for each individual formula and the "most probable" force derived as the average of the applicable formulas (upper boundary of the red shaded area). In the absence of calibration data for these measured profiles, a judgement can only be made on the general behaviour of each formula and on the results relative to each other. Chen (2015) and EurOtop were applicable most of the time, provided results within the range of average forces and intersected each other. Wave force calculations based on the Eurotop formula gave reasonable force estimates, in the sense that despite their theoretical character they matched with the empirical formula results of Chen (2012 and 2015) and were situated in the average of all formulas. Overall, results were characterized by large spatial variability with horizontal wave forces ranging from 0 to 100 kN/m, and the decrease of the force with distance from the crest.

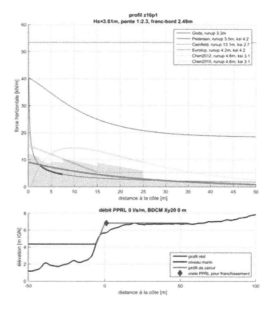

Figure 1. Calculated horizontal (F_h) force distribution along profile (top) and associated profile characteristics (bottom).

REFERENCES

Alp'Géorisques and IMDC. 2014. Plan de Prévention des Risques Littoraux en Vendée. Rapport de Phase 2.

Chen, X., Hofland, B., Altomare, C., Suzuki, T., and Uijttewaal, W. 2015. Forces on a vertical wall on a dike crest due to overtopping. *Coastal Engineering* 95, 94–104.

Chen, X., 2011. Hydrodynamic loads on buildings caused by overtopping waves. M.S. Thesis T.U. Delft, Netherlands; Project: 770_59 Flanders Hydraulics, Antwerp, Belgium.

Pullen, T., Allsop, N.W.H., Bruce, T., Kortenhaus, A., Schüttrumpf, H. & Van der Meer, J.W. 2007. EurOtop. Wave overtopping of sea defences and related structures.

U.S. Army Corps of Engineers. 2006. Coastal Engineering Manual. Man. 1110-2-1100, U.S.A.C.E, Washington, D.C.

Coastal tsunami-hazard mapping

Cuneyt Yavuz & Elcin Kentel
Department of Civil Engineering, METU, Ankara, Turkey

ABSTRACT: The Mediterranean Sea is approximately 3900 km long, its maximum width is around 1600 km and greatest depth is 4400 m. it Is one of the biggest marginal seas on the planet. During the last 36 centuries, 67 earthquakes with magnitudes greater than 7, 133 earthquakes with magnitudes between 6 and 7 and at least 96 tsunamis were documented in the Eastern Mediterranean (Altinok & Ersoy, 2000).

Mediterranean experienced numerous earthquakes and some of them triggered tsunamis throughout the history. Historical records, locations of fault zones and volcanoes indicate that there are some tsunamigenic sources in the Mediterranean Sea (see Figure 1).

For instance, 365 Crete Earthquake is one of the greatest tsunami incidents in Eastern Mediterranean. It is estimated that the magnitude of the earthquake was higher than 8.5 and was followed by a tsunami. Tsunami destroyed coasts of the East Mediterranean from Sicily to Egypt. Hundreds of thousands casualties were recorded and almost all the towns along the Eastern Mediterranean coasts were devastated (Carayannis, 2011).

Nowadays, tsunami hazard is still among the main projects of the European Union and the rest of the world. Thus, tsunami hazard assessment of the region is important for designing early warning systems, site selection of future critical infrastructures (CIs) and planning necessary mitigation measures for existing CIs and settlement areas.

Probabilistic tsunami assessments should be conducted to analyze likely adverse effects of tsunamis along coastlines using numerical modelling tools. Different numerical models have been developed for tsunami simulations since 1960's. Some of the most commonly used Tsunami modelling softwares are TUNAMI N1 (Imamura, 1995), MOST (Titov, 1999), NAMI DANCE (Zaytsev and Yalçıner, 2006) and SIFT (NOAA, 2015).

In this study, a probabilistic analysis of tsunamis is conducted along the Cyprus arc. Fitted probability density functions (PDFs) of the focal depth and the magnitude are used together with source parameters of historical earthquakes to generate PDFs of inundation depth at a number of CIs along the Gulf of Mersin. Locations of the simulated earthquakes are selected from the previously recorded tsunamis originated in the Mediterranean Sea and NAMI-DANCE software is used to estimate inundation depths. Selected CIs include ports, industrial zones, and waste water treatment plants. Simulated run-up results are assessed and PDFs of inundation depths at the CI locations are analyzed. It is concluded that some of the existing CIs may be vulnerable to probable future tsunamis.

Figure 1. Unit tsunami sources through the Hellenic and Cyprus Arcs using ComMIT interface.

REFERENCES

Altınok, Y. and Ersoy, Ş. 2000. Tsunamis observed on and near Turkish coast, *Natural Hazards, Vol. 21, No. 2–3, 185–205. Doi:10.1023/A:1008155117243*
Carayannis, P.G. 2011. The earthquake and tsunami of July 21, 365 ad in the eastern Mediterranean sea – Review of Impact on the Ancient World – Assessment of Recurrence and Future Impact, *Science of Tsunami Hazards, Vol. 30, No. 4, page 254. ISSN 8755-6839.*
Community Model Interface for Tsunami (ComMIT), *NOAA Center for Tsunami Research, PMEL.* Available at: http://nctr.pmel.noaa.gov/ComMIT/
NAMI DANCE, 2011, Manual of Numerical Code NAMI DANCE, *published in http://namidance.ce.metu.edu.tr*

Waves generated by ship convoy: Comparison of physical and numerical modeling with in-situ measurements

T. Cohen Liechti, G. De Cesare & A.J. Schleiss
Laboratory of Hydraulic Constructions (LCH), Ecole Polytechnique Fédérale de Lausanne (EPFL), Lausanne, Switzerland

R. Amacher
Laboratory of Applied Mechanics and Reliability Analysis (LMAF), Ecole Polytechnique Fédérale de Lausanne (EPFL), Lausanne, Switzerland

M. Pfister
Filière de génie civil, Haute école d'ingénierie et d'architecture (HEIA-FR), Fribourg, Switzerland

ABSTRACT

Since several decades, a part of the domestic waste of the City of Geneva (Switzerland) is transported on the Rhone River from the city center to the waste incineration station outside of the city with ship convoys consisting of a pusher tug and a barge. The pusher tug has a length of 12.1 m, a width of 5.5 m, and a weight of 52 tons; The barge is 43.0 m long, 8.6 m wide, and weighs 120 tons. The transport capacity of a barge is 170 t.

Waves generated by these convoys may damage the riverbanks and affect the riparian fauna (Nanson et al. 1994, Coops et al. 1996, Bishop 2003, De Roo et al. 2012), requiring consequently protection and maintenance measures. As an efficient approach, a reduction of the convoy velocity is frequently discussed, particularly as the convoy passes a nature reserve. The latter is, however, not appropriate for logistic reasons, so that adaptions on the hull are considered by the Industrial Services of Geneva (SIG) as operator. To specify such adaptions, and to quantify their effect, SIG assigned the Laboratory of Hydraulic Constructions (LCH) of Ecole Polytechnique Fédéral de Lausanne (EPFL) to propose related options and to validate them with physical and numerical model tests. The latter indicated that a flap mounted at the barge stern is effective (Fig. 1), as the waves generated between the pusher tug and the barge are critical. Those are significantly reduced by the flap, as the pusher tug and the barge are hydrodynamically linked. The flap is operated by hydraulic cylinders and lowered during journey, but lifted up in the port to facilitate manoeuvres.

SIG owns one pusher tug and four barges, of which one was equipped with the recommended flap. In order to verify its efficiency, SIG appointed LCH with in-situ measurements of the waves generated by two types of barges: (1) modified including the flap, and (2) original, not modified barge without flap (Amacher et al. 2015). To complete the investigation, LCH conducted

Figure 1. Sketch of pusher tug (left) and barge (right), with wave-reducing flap at barge stern.

additional numerical simulations of the prototype situation. The present case study allows thus to compare the wave characteristic derived from two classical engineering design tools: physical and numerical modeling. Furthermore, the latter results compared with the in-situ measurements, in order to assess the reliability of the models. Herein, the efficiency of the flap is thus not in the focus, but the comparison of the two modeling tools with the in-situ tests.

REFERENCES

Amacher, R., Cohen Liechti, T., Pfister, M., De Cesare, G., Schleiss, A.J. (2015). Wave reducing stern flap on ship convoy to protect riverbanks. *Naval Engineers Journal* 127(1), 95–102.

Bishop, M.J. (2003). Making waves: the effects of boat-wash on macrobenthic assemblage of estuaries. PhD *Thesis*, University of Sydney, A.

Coops, H., Geilen, N., Verheij, H.J., Boeters, R., van der Velde, G. (1996). "Interactions between waves, bank erosion and emergent vegetation: an experimental study in a wave tank." *Aquatic Botany*, 53(3–4), 187–198.

De Roo, S., Vanhaute, L., Troch, P. (2012). "Impact of ship waves on the sediment transport in a nature friendly bank protection." *River Flow* 2012, M. Murillo ed., Taylor & Francis, London.

Nanson, G.C., Krusenstierna, A., Bryant, E.A., Renilson, M.R. (1994). "Experimental measurements on river-bank erosion caused by boat-generated waves on the Gordon River, Tasmania." *Regulated Rivers: Research & Management*, 9(1), 1–14.

Example of wave impact on a residential house

D. Wüthrich, M. Pfister & A.J. Schleiss
Laboratoire de Constructions Hydrauliques (LCH), EPFL, Switzerland

ABSTRACT

In the past hydrodynamic waves such as dam-break waves, impulse waves and tsunamis were considered extremely rare events and most design codes do not present guidelines in this domain. Some recent catastrophic events with high casualties and significant damages showed to the whole world the importance of wave-resistant houses as a mean to save people lives. On coastal areas around the Indian Ocean and at the mountains close to dam sights, most of the buildings do not exceed three floors, resulting into low structures characterized by a rectangular shape.

In the context of a broader research project investigating in a laboratory environment the structural loading during wave impact, the hydraulic behavior of the process is here described and discussed. The project was based on an experimental approach and waves were produced using a vertical release technique. A known volume of water was dropped from an upper reservoir into a lower reservoir through three identical pipes; the propagation took place in a 14 m long and 1.4 m wide horizontal smooth channel. Given their different behavior, both dry bed surges and wet bed bores were tested. The structure was simulated using a 30×30 cm aluminum cube, reproducing a residential house of 9×9 m if a Froude scaling ratio of 1:30 is assumed. For the same scaling ratio, wave heights ranged from 4.5 to 7.5 m, which is consistent with field observations.

When a building was hit by an incoming wave, the flow was 3-dimensional, highly turbulent and many phenomena appeared in a short period of time. Some visual observations were carried out using high-speed cameras to provide a better understanding of the time development of these phenomena. Visual observations

Figure 1. Impact of a dry bed surge against the aluminum structure representing a common residential house found in areas subject to tsunami hazard.

proved that overall surges and bores with the same impoundment depths presented similar behaviors. For all scenarios the wave impact presented high splashes and some turbulent air entrainment on the upstream side of the building. In general bores showed higher splashes, probably due to their steeper front. The presence of the building provoked a constriction, with a decrease in flow velocity and an increase of the water level. The combination of these effects lead to a change in flow regime and a propagation of a bore in the upstream direction. For all biggest waves an overflow was observed and the structure was fully submerged. For the smallest waves the structure was not overflowed and the roof could be used as a vertical shelter in case of tsunami alert.

Innovation and sustainability in hydraulic engineering and water resources

Estimation of rainfall-runoff relation using HEC-HMS for a basin in Turkey

H. Akay & M. Baduna Koçyiğit
Civil Engineering Department, Gazi University, Ankara, Turkey

A.M. Yanmaz
Civil Engineering Department, Orta Dogu Teknik Universitesi, Ankara, Turkey

ABSTRACT

It is crucial to foreknow the runoff hydrograph of a watershed for a precipitation pattern in order to estimate the maximum discharge required in design and operation stages of a hydraulic structure on rivers. The response of the watershed to a rainfall is affected by hydrologic processes which could be altered by the environmental factors, such as land use – land cover and global warming etc. Changing environmental conditions reflect the changing response of the nature. Therefore, the more accurate the hydrologic parameters are, the more accurate the hydrologic models can estimate the corresponding event based runoff hydrograph.

In this study, HEC-HMS is used to predict runoff hydrograph for a watershed located in the Western Black Sea Region Basin which has 30000 km^2 drainage area, having 811 mm of annual average rainfall with 9.93 km^3 flow volume is chosen as the study area (Yanmaz 2006). Kocanaz Creek which is a branch of the River Bartın passes through this sub basin. The basin collects water from an altitude of 1755 m and is very steep with an average slope of about 40% where the gradient exceeds 360% at some locations. Due to this steep topography, sudden flooding is quite common in the area. The study area in this research is quite steep with forested mountainous regions. There are also agricultural areas, and pastures. Hydrological soil groups in the basin are determined as C and D with an average curve number of about 78.5, this value typically ranging from 70 to 98 locally. These values show that major part of the rainfall continues as runoff in the watershed. Kocanaz Creek collects water from the area of Ulus town and Bartın city and passes through close to Bartın city center where Kocanaz stream gaging station is located which is taken as outlet of the basin.

HEC-HMS is first calibrated for a storm event occurred in 27 January – 7 February 2007. The model is then validated for another storm event in 2–10 March 2007.

Calibration and validation is performed for two successive storm events subdividing the watershed into seven and eleven sub basins. It was found out that increasing the number of division of the watershed improved the simulation results for calibration. This can be explained that when nearby sub basins with similar hydrologic behavior are grouped, the results can be improved. However, using successive storm events for validation resulted in underestimated values because of not predicting the CN values accurately. This can be explained by soil being saturated from first storm event. These are the preliminary results of an ongoing research in which such hydrologic simulations are performed in the area.

REFERENCE

Yanmaz, M. 2006. *Applied Water Resources Engineering.* Ankara: METU PRESS.

Using ANN and ANFIS Models for simulating and predicting Groundwater Level Fluctuations in the Miandarband Plain, Iran

M. Zare & M. Koch
Department of Geotechnology and Geohydraulics, University of Kassel, Kassel, Germany

ABSTRACT

The Miandarband plain is one of the most fertile plains of the Kermanshah province, Iran. The major water supply for agriculture is groundwater. In this regard, simulation and prediction of groundwater level (GL) fluctuations plays an important role for effective water resources management. GL-changes are complex to model, as they depend on many nonlinear and uncertain factors, thus selecting suitable numerical or stochastic models that could simulate the nonlinearity and complex patterns is of great importance. Artificial Neural Networks (ANN) and/or fuzzy logic models are one family of models that have proven to be very useful to that regard. In this study, after data completion, the Feed Forward Neural Network (FFNN) model with one hidden layer whose perceptrons have been optimized in the -training phase with three methods (Levenberg-Marquardt, Bayesian regularization and Scaled Conjugate Gradient). From the statistical analysis the optimal number of neuron in the hidden layer of the FFNN models is determined as 10, with a minimum RMSE of 0.61, i.e. a neural network model with a (7-10-1) structure has been applied in the present study.

The Adaptive Network-based Fuzzy Inference System (ANFIS) has been applied and evaluated for GL-fluctuations simulation and prediction in the Miandarband plain, Iran. In this regard, a fuzzy rule generation technique that integrates ANFIS with the Fuzzy Clustering Method (FCM) is applied to minimize the number of fuzzy rules. The results show that both model approaches can be used with acceptable accuracy, wherefore the ANFIS- model is better than the FFNN- models. The conclusions achieved from this study are that the ANFIS-system can be considered as a good data-driven model for the prediction of groundwater level fluctuations due to recharge and discharge process in an aquifer and, consequently, that it can be used for planning of sustainable yields of groundwater resources.

Table 1. Results of models in the training phase.

Model	RMSE	R^2
FFNN-LM	0.61	0.965
FFNN-BR	0.68	0.939
FFNN-SCG	0.71	0.927
ANFIS	**0.48**	**0.967**

Table 2. Results of models in the testing phase.

Model	RMSE	R^2
FFNN-LM	0.88	0.920
FFNN-BR	0.98	0.875
FFNN-SCG	0.69	0.930
ANFIS	**0.52**	**0.962**

REFERENCES

Daliakopoulosa, I. N., Coulibalya, P. and Tsanisb, I. K. 2005. Groundwater level forecasting using artificial neural networks. Journal of Hydrology 309. 229–240.

Shirmohammadi B, Vafakhah M, Moosavi V, Moghaddamnia A., 2013. Application of several data-driven techniques for predicting groundwater level. Water Resour Manag., 27, 2, 419–432.

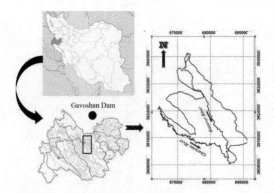

Figure 1. Location of the Miandarband plain, Iran.

A numerical groundwater flow model of Bursa Basköy aquifer

S. Korkmaz & G.E. Türkkan
Department of Civil Engineering, Uludag University, Bursa, Turkey

ABSTRACT

In this study, the aim is to determine the groundwater storage and distribution in the Basköy aquifer inside the city of Bursa, Turkey. Preliminary studies were conducted in the GIS environment, and afterwards, groundwater flow simulations were performed in MODFLOW which uses finite differences method to solve groundwater flow equations. From point elevation data, a digital elevation model (DEM) was constructed. The subbasins and drainage network were delineated through hydrological analysis of the DEM. The data were transferred to MODFLOW interface and a grid with a cell size of 15 m × 15 m was generated. Hourly groundwater level measurements at 7 observation wells were obtained from State Hydraulic Works. The observation data were used for calibration of unknown parameters such as hydraulic conductivity and specific yield.

Two types of simulations were performed, namely, steady-state and transient. The former was used to obtain a general groundwater head distribution and rough values of hydraulic conductivity by using annual mean of rainfall data. Calibration process was expedited by using the Parameter Estimation (PEST) model. When using this model, aquifer is divided into parameter zones. In a parameter zone, the value of a calibration parameter is the same in all of the cells. Initially subbasins were used as parameter zones, and in later stages number of zones were increased in order to have a more heterogeneous distribution. PEST tries to find the optimum value of parameters using observation data.

In transient simulations, EVT (evapotranspiration) package was also enabled and daily evapotranspiration values were used. Simulations were performed in daily timesteps between 1/12/2013 and 31/12/2014. According to results, model output was dependent on the initial head distribution, hydraulic conductivity and storage coefficient. In most of the wells general groundwater dynamics were successfully predicted, however, the sharp peaks in observations are yet to be predicted. They are thought to be due to the karstic nature of the aquifer.

REFERENCES

Doherty, J., (2013). PEST: Model-Independent Parameter Estimation. Watermark Numerical Computing, Brisbane, Australia.

Harbaugh, A.W. (2005). MODFLOW-2005, The U.S. Geological Survey modular ground-water model—the Ground-Water Flow Process: U.S. Geological Survey Techniques and Methods 6-A16. Available at: http://pubs.water.usgs.gov/tm6a16

McDonald, M.G., Harbaugh, A.W., (1988). A modular three-dimensional finite-difference ground-water flow model: Techniques of Water-Resources Investigations of the United States Geological Survey, Book 6, Chapter A1, 586 p.

Niswonger, R.G., Panday, S., Ibaraki, M., (2011). MODFLOW-NWT, A Newton formulation for MODFLOW-2005: U.S. Geological Survey Techniques and Methods 6–A37, 44 p.

SHW (2013). Bursa-Orhaneli Baþköy-Ortaköy Kuzeyi Kireçtaþları Karst Hidrojeolojik Etüt Raporu. General Directorate of State Hydraulic Works. [In Turkish]

Groundwater management and potential climate change impacts on Oum Er Rbia basin, Morocco

M. El Azhari & D. Loudyi
Water and Environmental Engineering, University of Hassan II Mohammedia–Casablanca, Mohammedia, Morocco

ABSTRACT

The aquifers of Oum Er Rbia basin show an imbalance between the input and output because of their overexploitation for agriculture and drinking water. Indeed, in 2012, groundwater extractions reached 608 million m^3 while groundwater potential was only 347 million m^3, yielding a deficit of nearly 300 million m^3.

Moreover, climate change will have a certain impact on the future of these resources. Further, a study conducted in 2013 by the Hydraulic Basin Agency of Oum er Rabia and the World Bank, predicted increases in the mean annual temperature on the order of 0.1 and 1.4°C by the period from 2010 to 2030. In addition, the average annual rainfall will decrease by about 200 mm. The demand for irrigation water and drinking water will however increase given the population growth coupled with economic development. Climate change will adversely impact aquifers recharge and the level of their water table, and consequently, the gap between supply and demand will still grow.

Modeling groundwater resources taking into account regional climate scenarios is necessary to predict the potential impacts of climate change on groundwater resources sustainability.

In this work, a reflection on a more efficient and sustainable management of groundwater resources in the Oum Er Rbia basin will be presented. The primary goal is to fill the gap between supply and demand through climate change adaptation actions, in particular, the use of non-conventional water resources such as desalination of sea water, reuse of treated wastewater, rain water harvesting, optimization of irrigation infrastructures and the adoption of projects of water transfer from other national basins with excess water.

REFERENCES

River Basin Agency of Oum Er Rbia, 2012. Impacts of Global Climate Change (GCC) on the Water Resources of Morocco, Phase 2: Hydrological Modeling for Oum Er Rbia River Basin and Phase 3: Integrated Water Resources Simulation for Oum Er Rbia River Basin.

River Basin Agency of Oum Er Rbia, 2012. Master Plan of Integrated Water Resources Management (MPIWRM).

Flow modeling in vegetated rivers

M. Mabrouka
National Institute of Agronomy of Tunisia, University of Carthage, Tunis, Tunisia

ABSTRACT

The development of vegetation in river may increase flood risks. It plays an important role in influencing the hydrodynamic behavior, ecological equilibrium and environmental characteristics of water bodies. Therefore, it's important to have suitable prediction of increased resistance caused by vegetation. In the recent years, experimental and numerical models have both been developed to model the effects of submerged vegetation on open-channel flow. This paper describes a new analytic model based on physical equations used to predict hydraulic conditions in vegetated rivers. This model was programmed in two-dimensional software to predict flow characteristic in vegetated flow. Simulations were applied on river reach of Medjerda in Tunisia. The comparison between simulated and observed results shows a good performance of this model in the prediction of flood in vegetated rivers.

Comparison of methods to calculate the shear velocity in unsteady flows

G. Bombar
Ege University, İzmir, Turkey

ABSTRACT

The shear velocity is an important parameter in all areas of hydraulics, therefore it is essential to calculate it correctly. There exist methods to estimate the shear velocity in steady flows, but the application of these methods to unsteady flows is limited. In this study, an artificial triangular-shaped hydrographs was generated in a rectangular flume of 70 cm wide and 18 m long.

The time series of the shear velocity was obtained by several methods, such as, u_{*SV} by using the Saint-Venant equations, u_{*L} by using the procedure given by Clauser Method, u_{*P} by using the parabolic law, u_{*UN} by using the momentum equation assuming the slope of energy grade line is equal to bed slope and u_{*avg} by using the average velocity equation.

In order to examine the depth variation of hysteretic behavior of depth variation with point velocity, a hysteresis intensity parameter η was used. Looking at the variation of η with normalized elevation as z/h_{base}, one can observe that there is a relation between these two dimensionless parameters as given below for H1.

$$\eta = a \left(\frac{z}{h_{base}}\right)^b \quad (1)$$

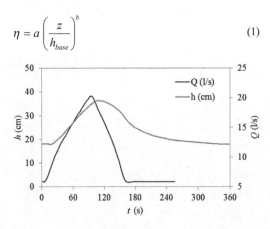

Figure 1. Flow depth and discharge variation of experiment H1 with time.

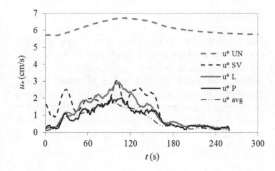

Figure 2. Variation of shear velocities u_{*UN}, u_{*SV}, u_{*L}, u_{*P} and u_{*avg} for experiment H1 with time.

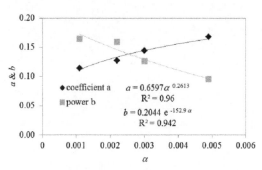

Figure 3. Variation of coefficient and power in Eq. (1) with α.

Examining the data of Bombar (2016), it is observed that the coefficient a and power b in Eq. (1) depends on the unsteadiness parameter α.

REFERENCE

Bombar, G. (2016). The Hysteresis and Shear Ve-locity in Unsteady Flows. *Journal of Applied Fluid Mechanics* 9(2), 839–853.

Non-linear optimization of a 1-D shallow water model and integration into Simulink for operational use

L. Goffin, B.J. Dewals, S. Erpicum, M. Pirotton & P. Archambeau
HECE, University of Liege (ULg), Belgium

ABSTRACT

Dam operations leads to discharge changes in the downstream river. Evaluating the impact of such changes can help to improve the management of these hydraulic structures. This can be achieved thanks to a well calibrated 1-D model which offers a very low computational cost compared to 2-D models. In this paper, WOLF-1D is used. It implements shallow water equations discretized in 1-D finite volumes. All presented developments are applied to the practical case of the River Romanche, in the French Alps.

The 1-D model accuracy is mainly a function of the friction representation and the topography. In order to get best results, an original technique is proposed to generate 1-D cross sections from a set of existing 2-D stationary simulations at various discharges. For each simulation and each 1-D node, the bottom altitude, the cross section area and wet perimeter can be determined. This leads to tabular relations for each node. The Barr-Bathurst law (Machiels et al. 2011) was used for friction.

Getting the 1-D model as close as possible to the 2-D model is achieved by calibrating some parameters. A first possibility is to calibrate the roughness coefficient. It can be distributed in different zones of the domain for more flexibility. The calibration was done by scanning a predefined space of roughness coefficients. This technique can be efficient when few parameters should be calibrated.

As a second step, a tuning coefficient was added to the computation of the wet perimeter χ:

$$\chi = \chi_{tr}\left(\alpha\left(\frac{h}{h_{max}}\right)+1\right)$$

This formula is based on the wet perimeter saved in the tabular relations χ_{tv}, the current water height compared to the maximum level recorded in 2-D h/h_{max} and the tuning factor α. Simulated annealing method (Corana et al. 1987; Goffe et al. 1994; Kirkpatrick et al. 1983; Metropolis et al. 1953), a heuristic optimization method, was used for the calibration of two parameters, distributed in two zones: roughness sizes and wet perimeter coefficients. Comparing results obtained with both optimization methods showed that the second technique offers better agreement with reference data.

Integrating computational tools into already operational systems is a major concern for hydraulic facilities manager. Simulink offers the possibility to integrate routines coded in C or FORTRAN. This is done thanks to S-Functions which require an input port as well as an output one. Inputs can be, for example, upstream hydrographs and lateral discharges while the output can produce the water level evolution at a given point or a downstream hydrograph.

Application of these developments on the River Romanche showed encouraging results.

REFERENCES

Corana, A. et al., 1987. Minimizing multimodal functions of continuous variables with the "simulated annealing" algorithm. ACM Transactions on Mathematical Software, 13(3), pp. 262–280.

Goffe, W.L., Ferrier, G.D. & Rogers, J., 1994. Global optimization of statistical functions with simulated annealing. Journal of Econometrics, 60(1), pp. 65–99.

Kirkpatrick, S., Gelatt, C.D. & Vecchi, M.P., 1983. Optimization by simulated annealing. Science, 220(4598), pp. 671–680.

Machiels, O. et al., 2011. Theoretical and numerical analysis of the influence of the bottom friction formulation in free surface flow modelling. Water SA, 37(2), pp. 221–228.

Metropolis, N. et al., 1953. Equation of state calculations by fast computing machines. The journal of chemical physics, 21(6), pp. 1087–1092.

Limitation of self-organization within a confined aquifer

M.C. Westhoff, S. Erpicum, P. Archambeau, M. Pirotton & B. Dewals
Hydraulics in Environmental and Civil Engineering (HECE), University of Liege (ULg), Liege, Belgium

E. Zehe
Karlsruhe Institute of Technology (KIT), Karlsruhe, Germany

ABSTRACT

Preferential flow paths are omnipresent in the subsurface, but very hard to observe or parameterize. Often they are even sub-grid processes requiring effective parametrization. Since such effective parameters at the scale of the model are generally not available, current practice is to calibrate model parameters, resulting in an ill-posed problem with many parameter combinations revealing the same results.

To overcome this calibration paradigm, self-orginzing principles have been proposed. Such principles should be able to explain the current behaviour of a system. One of such principles is the maximum power principle, stating that a system organizes itself such that the flux of interest produces maximum power.

In an ongoing study, Westhoff et al. (2016) show within a lab experiment that the steady state effective hydraulic conductance evolves to the conductance that maximizes power by the flux through the confined aquifer. In their case, the effective hydraulic conductance could evolve by creating preferential flowpaths caused by internal erosion of sand particles, aka piping.

However, the maximum power principle could only succesfully predict the effective conductance in some sets of boundary conditions, while the principle was unable to predict effective conductance of some others sets of boundary conditions.

In this study we explore why in one setup the effective conductance did obey this maximum power principle, while the same setup with slightly different boundary conditions did not lead to this. We hypothesized that the presence of a high-conducting foam rubber layer on top of the confined aquifer limited the maximum length of the evolving preferential flowpath and that this maximum length was long enough for one set of boundary conditions and appeared to be to short for the other.

We "confirmed" this hypothesis with a detailed 3D numerical model (WOLF 3D) of the experiment. Without the high-conducting foam rubber layer, the pipe would evolve as in Fig. 1a, while with the foam rubber layer, the pipe evloves as in Fig. 1b. By only varying the

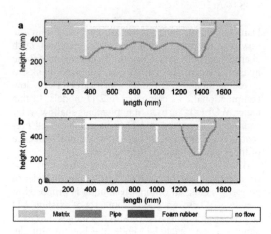

Figure 1. 2-D slice of pipe layout in WOLF 3D for a) configuration without foam rubber and b) configuration with foam rubber layer.

conductivity of the pipe elements, a theoretical (modeled) maximum in power could be found in the former case for both sets of boundary conditions, while in the latter setup, a maximum could only be found for one set of boundary conditions. So we conclude that most likely, the degrees of freedom to let a preferential flow path evolve was too limited for the setup for which the effective hydraulic conductance could not be predicted by the maximum power principle.

REFERENCE

Westhoff, M., S. Erpicum, P. Archambeau, M. Pirotton, E. Zehe, & B. Dewals (2016). Experimental proof that the effective soil hydraulic conductivity can be predicted with the maximum power principle. in preparation.

Preliminary experiments on the evolution of river dunes

G. Oliveto & M.C. Marino
School of Engineering, University of Basilicata, Potenza, Italy

ABSTRACT

The hydraulic roughness of rivers depends greatly on the bedform configuration. Dunes are common bedforms that typically develop in beds with sediment sizes from medium sand to gravel. They tend to migrate downstream and are out of phase with the water surface waves. Their longitudinal profiles are often asymmetrical with fairly mild stoss side slopes (around 5°) and steeper lee side slopes close to the angle of repose of the bed material (Julien 2010).

This paper is a part of ongoing investigations by the authors on the evolution of river dunes. Here the spatial and temporal development of a solitary sand dune is investigated experimentally.

Experiments were carried out at University of Basilicata, Italy, in a 1 m wide and 20 m long rectangular channel. A nearly-uniform sand with median grain size $d_{50} = 1.7$ mm was used as mobile bed. The working section was 16 m long. Runs were performed in two phases. Phase 1 aimed to generate steady-state dune shapes starting from an initial arc-shaped dune, of the same material as the mobile bed, with the axis perpendicular to the longitudinal channel axis. Phase 2 aimed to promote the dune propagation starting from the natural-like dune generated in Phase 1 by lowering the tailwater level, but keeping constant the discharge. Phase 2 would then simulate the dune morphology evolution under unsteady flow conditions when flow depth decreases but discharge remains the same. Figure 1 shows the development of the starting arc-shaped dune towards a more natural dune form in case of run A1. A diagram with the axial bed profile is associated with each photograph to better appreciate the dune geometrical characteristics. In general the starting symmetrical arc-shaped profile evolved into an asymmetrical profile with a gently inclined stoss side, which slope was somewhat smaller than 5° and a steeper lee face, which slope was found dependent on the approach Froude number and not strictly corresponding to the angle of repose of the bed sediment.

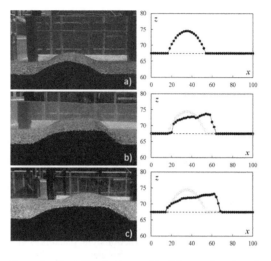

Figure 1. Evolution during run A1 of the arc-shaped sand dune towards its steady-state shape. Photographs correspond to times t from run starting (a) $t = 0$ h, (b) $t = 0.1$ h, and (c) $t = 1.5$ h. z is the bed elevation (in cm) and x is the axial (longitudinal) coordinate (in cm). Gray symbols represent the starting dune configuration.

Based on the experimental data new insights are provided on equilibrium morphology and rate of migration of solitary sand dunes. Also a 2D numerical model was applied to simulate the evolution of solitary dunes for runs of Phases 1 and 2. Results were found in satisfactory agreement with experimental observations in the case of the dune height and also in the case of the dune length for runs of Phase 2. Conversely, significant discrepancies were identified in the simulation of the dune shape development.

REFERENCE

Julien, P.Y. 2010. *Erosion and sedimentation* – 2nd Edition. Cambridge: Cambridge University Press.

A conceptual sediment transport simulator based on the particle size distribution

B.T. Woldegiorgis, W. Bauwens & M. Chen
Department of Hydrology and Hydraulic Engineering, Vrije Universiteit Brussel, Belgium

F. Pereira
Flanders Hydraulics Research (Waterbouwkundig Laboratorium), Belgium

A. van Griensven
Department of Hydrology and Hydraulic Engineering, Vrije Universiteit Brussel, Belgium
Chair Group of Hydrology and Water Resources, UNESCO-IHE Institute for Water Education, The Netherlands

ABSTRACT

Sediment transport is a very important process affecting the transport and retention of pollutants that easily adsorb to suspended sediments. The fine sediments play a dominant role and need to be properly represented. However, the conventional conceptual sediment transport models do not account for the distribution of the particle sizes and the differences in behaviour of the different fractions. Our paper presents an analytic solution for this problem, using the log-normal probability density function to represent the particle size distribution of the sediment. Sedimentation and resuspension processes are calculated by considering the particle size distributions and integrating Velikanov's energy equation with the Hjulstrom diagram, as recently modified by Miedema. The new method imposes the critical condition of deposition after checking the maximum carrying capacity and applies the critical condition of initiation of motion before imposing the minimum carrying capacity.

An application on the Zenne River (Belgium) shows that our conceptual model provides a better representation of the high concentrations and range of concentration than the conventional conceptual sediment transport modelling. The method is able to mobilize deposition of sediment particles coarser than the critical diameter even if sediment concentration in the channel does not violate the maximum carrying capacity of the stream. It also adapts the settling velocity to the changing median particle size and hence determines a more realistic sediment carrying capacity. This allowed for a better simulation of the extreme low and high sediment concentrations, which the conventional conceptual sediment transport models could not handle.

REFERENCES

Miedema, S., 2010. Constructing the Shields curve, a new theoretical approach and its applications, in: WODCON XIX. WODCON XIX, Beijing China.

Velikanov, M.A., 1954. Gravitational Theory of Sediment Transport. J. Sci. Sov. Union, Geophys. Russ. 4.

Effect of seepage on the friction factor in an alluvial channel

P. Mahesh, V. Deshpande & B. Kumar
Department of Civil Engineering, Indian Institute of Technology Guwahati, India

ABSTRACT

Alluvial channels often encounter permeable boundaries in natural environments such as porous boundaries consisting of sediment particles in natural rivers and in irrigation canals. Exchange of water (seepage) can occur in either way, flow from the channel (downward seepage) or into the channel, depending upon the difference of level between the water in the channel and the surrounding groundwater table.

However, understanding of the interaction between groundwater and surface water is necessary in order to find its influence on the friction factor. Thus, the objective of this study is to investigate the friction factor over sand bed channel in the presence of downward seepage. Further, the present work also focuses on the variation in the value of von-Karman's constant in alluvial channel under the action of downward seepage.

In this study, an experimental approach is used to observe the effects of downward seepage on flow resistance in a threshold alluvial channel with parabolic cross sectional shape. Experiments were performed in a 20 m long, 1 m wide, and 0.72 m deep recirculating transparent plexi-glassed tilting flume. They were carried out on two uniform river sands: fine grained ($d_{50} = 0.41$ mm) and coarse grained ($d_{50} = 1.1$ mm). To analyze the turbulent flow statistics, instantaneous flow measurements were taken with the help of Vectrino$^+$ Acoustic Doppler Velocimeter developed by Nortek. Samples were collected for five minutes duration at a sampling rate of 100 Hz. In order to understand the flow hydrodynamics, turbulent flow parameters such as time-mean velocities, Reynolds shear stress, and shear velocity are analyzed for no seepage and seepage experiment.

We have observed that Reynolds shear stresses are increased for seepage experiment (immediately after the application of seepage), suggesting an increased momentum transfer toward the channel bed under the action of downward seepage. In turbulent flows, shear velocity defines the scale for fluctuating component of the velocity. The values of shear velocities are evaluated as 0.016 m/s (0.023 m/s) and 0.019 m/s (0.026 m/s) over fine (coarse) sand bed for no seepage and seepage experiment, respectively. A higher value of shear velocity for seepage experiment indicates that scales of velocity fluctuations in the flow are increased in the presence of downward seepage.

We can observe that the universal logarithmic law of wall for flows over rough boundaries does not apply for the flow over parabolic cross-sectional sand bed. Equation of the universal logarithmic law ($u^+ = 2.43 * \ln(z^+) + 8.5$) has been modified for no seepage experiment, considering the flow in a channel with curved cross-section. Because of the strong exchange of momentum between the flow over the channel bed and the seepage flow, the modified logarithmic law for no seepage experiment had to be further modified considering the downward seepage scenario.

It has been observed that for the no seepage experiment, value of von-Karman's constant agrees well with the universal value. However, for the seepage experiment the value of von-Karman's constant is reduced. Reduction in the value of k is in agreement with the increased sediment transport, which occurred in the channel after the application of downward seepage.

In order to determine the friction factor in flow, Darcy-Weisbach relation is used. The values for friction factor are calculated as 0.024 (0.039) and 0.031 (0.047) for no seepage and seepage experiment over fine (coarse) sand bed channel, respectively. This increase in the value of friction factor after the application of downward seepage indicates that downward seepage causes higher roughness on the channel bed and the resistance to flow is significantly increased under the condition of downward seepage.

Furthermore, We have analyzed the turbulent production (T_P) and turbulent kinetic energy dissipation (E_D) for experiments on both sands under no seepage and seepage conditions. Is has been observed that the magnitude of T_P is increased when the downward seepage is applied to the channel. This increased T_P is in agreement with the higher momentum transfer under the seepage condition. Also, it can be implied that larger roughness over the boundary of the channel with the application of seepage, causes more turbulent production in the flow. Also, result shows that the E_D is reduced when water is extracted in the downward direction from the sand bed. Under the action of downward seepage, more energy is converted to turbulent fluctuations in the flow, resulting in the reduced dissipation of the turbulent kinetic energy.

Experimental investigation of light particles transport in a tidal bore generated in a flume

L. Thomas, W. Reichl, Y. Devaux, A. Beaudouin & L. David
Axe HydEE, Département D2, Institut Pprime,
Université de Poitiers · CNRS · ISAE-ENSMA, Poitiers, France

ABSTRACT

Tidal bores are natural flows that can have a strong impact on river ecosystems. Recently, there has been a regain of interest in this flow, for example in China. Many in-situ measurement campaigns have been performed, but they are very limited concerning the spatial resolution. Concerning the impact of the tidal bore on the sediment transport, see for example Chen (2003), Chanson (2005), Simon et al. (2011) or Fan et al. (2012). Studies in the lab inside flumes, although questionable concerning their similitude to real tidal bores, can give new insights to understand the impact of such phenomena on the river sediments. In this study, optical measurement techniques (stereo-PIV and PTV coupled) are used to add some understanding of the influence of moving hydraulic jumps on the sediment transport. The experiment is performed in a

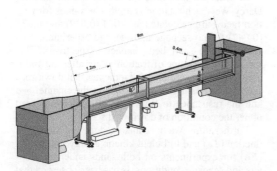

Figure 1. Sketch of the experimental setup.

Figure 2. Visualizations of the trajectories under the bore front for Fr = 1.58. Each line has a duration of 0.16 s.

Figure 3. FTLE field under the bore front for Fr = 1.58. The integration time is $T = 0.1$ s. The highest values correspond to the red color, and the lowest to the blue color.

flume (see figure 1). Four Froude numbers are investigated. It is shown that the particles follow the flow without significant differences. The trajectories (see example in figure 2) and their stability (see example in figure 3) are observed using sliding averages of the particle images and Finite Time Lyapunov Exponents (FTLE). It is shown that for undular bores, the impact on the trajectories is weak, while strong effects are observed in the breaking bore case.

ACKNOWLEDGEMENT

The authors would like to acknowledge the French National Agency for their financial support in the Project BLANC MASCARET 10-BLAN-0911, the FEDER and the Region Poitou-Charentes.

REFERENCES

Chanson, H. (2005). Physical modelling of the flow field in an undular tidal bore. *Journal of Hydraulics Research 43(3)*, 234–244.
Chen, S. (2003). Tidal bore in the north branch of the changjiang estuary on erosion and sedimentation. In *Proc. International Conference on Estuaries and Coasts 1*, Hangzhou, China, pp. 233–239.
Fan, D., G. Cai, S. Shang, Y. Wu, Y. Zhang, & L. Gao (2012). Sedimentation processes and sedimentary characteristics of tidal bores along the north bank of the qiantang estuary. *Chinese Science Bulletin 57*, 1478–1589.
Simon, B., P. Lubin, S. Glockner, & H. Chanson (2011). Three-dimensional numerical simulation of the hydrodynamics generated by a weak breaking tidal bore. In *Proc. 34th IAHR World Congress*, Brisbane, Australia.

Effect of hydrodynamics factors on flocculation processes in estuaries

A. Mhashhash, B. Bockelmann-Evans & S. Pan
School of Engineering, Cardiff University, UK

ABSTRACT

Cohesive sediment is primarily composed of clay minerals and organic matter (Manning et al. 2004), and has an ability to flocculate into large aggregates, namely flocs, which are bigger than individual particles but less dense. According to previous flocculation studies, flocs are classified according to their size into micorflocs, which have a spherically equivalent diameter of 160 μm with a settling velocity of less than 1 mm/s; and macroflocs which are form by combining mairoflocs with a diameter of more than 160 μm and settling velocities ranging from 1–15 mm/s (Eisma 1986; Fennessy et al. 1994; Manning and Dyer 1999; Manning 2001; Manning et al. 2013). The flocculation process mainly occurs in the very low salinity region (between 1 and 2.5 ppt) (Wollast 1988), and it is affected by the changes of hydrodynamic conditions which can alter the suspended sediment particles by modifying its effective particle size, shape, porosity, density and composition (Wollast 1988; Dyer 1994). In this study, laboratory experiments were performed to investigate the effect of turbulence on floc size and settling velocity. Experiments were conducted in a 1 L glass beaker of 11 cm diameter using suspended sediment samples taken from the Severn Estuary. A non-laser based Particle Image Velocimetry (PIV) system, as shown in Figure 1, and an advanced image processing software were used to measure floc size distribution and settling velocity. It was found that the settling velocity for the sediment used in the experiments ranged from 0.4 to 1.35 mm/s. The average settling velocity increased with turbulence up to a maximum value (1.1 mm/s) above which it decreased again. It was also found that the turbulent shear stress (T_s) level ranging from 0.57 to 8.5 N/m^2 causes a breakdown in floc structure rather than enhancing the flocculation process. The relationship between floc size and turbulence was found to be independent of the history of the floc formation.

REFERENCES

Dyer, K. H. 1994. Estuarine sediment transport. In: PYE, K. (ed.), Sediment transport and depositional processes. Blackwell. Scientific Publications, pp. 193–218.

Eisma, D. 1986. Flocculation and de-flocculation of suspended matter in estuaries. Netherlands Journal of Sea Research 20(2-3), pp. 183–199.

Fennessy, M. J. et al. 1994. inssev: An instrument to measure the size and settling velocity of flocs in situ. Marine Geology 117(1–4), pp. 107–117.

Manning, A. J. 2001. Study of the effect of turbulence on the properties of flocculated mud University of Plymouth.

Manning, A. J. and Dyer, K. R. 1999. A laboratory examination of floc characteristics with regard to turbulent shearing. Marine Geology 160(1–2), pp. 147–170.

Manning, A. J. et al. 2004. Flocculation measured by video based instruments in the gironde estuary during the European commission SWAMIEE project, Journal of Coastal Research (SPEC. ISS. 41), pp. 58–69.

Manning, A. J. et al. 2013. Flocculation Dynamics of Mud: Sand Mixed Suspensions.

Wollast, R. 1988. The Scheldt Estuary. W. Salomon, B.L. Bayne, E.K. Duursma, U. Förstner (Eds.), Pollution of the North Sea, an assessment, Springer-Verlag, pp. 183–193.

Figure 1. Laboratory setup for settling and flocculation measurements.

Lack of scale separation in granular flows driven by gravity

G. Rossi & A. Armanini
Department of Civil, Environmental and Mechanical Engineering, University of Trento, Trento, Italy

ABSTRACT

Since the 80s, kinetic theories of gases have been applied to granular flows: the collisions among particles are considered the equivalent of molecules collisions. In particular the thermodinamic temperature considered for gases, is substituted by a kinetic temperature in granular flows: it represents the kinetic energy of the particles ensamble and depends on the velocity fluctuations.

There are important differences between the two applications: in molecular gases the microscale, represented by the mean free paths, is very much smaller than the macroscale, that is the scale according to which gradients change (Goldhirsch 2008), while in granular flows driven by gravity the dimension of a single particle becomes comparable to that of the control volume. In gases the number density (which represents the number of particle in a certain volume) tends to inifinity, so that it can be considered constant, but in the granular case it is no more true. Because of this lack of scale separation, the volumes analyzed don't have a very large number of particles and even a few particles may change significantly the number density n, which can not be considered constant in more than one realization.

In the literature two main length scales have been identified in granular flows (Hsu et al., 2003): a small scale proportional to the particle size and a large scale related to the large eddies in turbulence. We introduce a third intermediate scale for granular flows, in order to solve the lack of scale separation.

A deep analysis of the implications due to the averaging processes applied at each scale must be taken into account. In this respect, we may consider two types of averages: the phasic average and the mass-weighted average (Drew 1983). In the case of pure fluid or of two phase fluid but with a constant number density, these two averaging processes coincide. This is not true in the case of granular flows composed by particles that cannot be considered small enough.

Furthermore, granular flows are not ergodic system (Jaeger et al. 1996): the average of a process parameter over time and over a stitistical ensemble are not the same. Since in the usual kinetic theory the esam-ble average is applied, particular attention should be paid to this issue.

For these reasons, the choice of a proper averaging process is crucial.

We believe that the intermediate scale previously mentioned could be solved applying a different kind of average, which accounts for the fluctuations of the number density and leads to new diffusive terms. The measurements done in a laboratory channel of the phasic-averaged vertical velocity and transversal velocity of the flow prove our hypotheses.

A diffusive coefficient will be introduced, whose scales are the square root of the granular temperature and the flow depth h or the particles distance from the wall (instead of the particle diameter d, used for the small scale). In this way, the closure relation of the small scale currently adopted in the kinetic theories will be extended to the intermediate scale. A proper definition of the diffusive coefficient will be provided through an intensive experimental investigation.

REFERENCES

Drew, D.A., 1983. Mathematical modelling of two-phase flow. Annu. Rev. Fluid Mech., 15, 261–291.
Goldhirsch, I., 2003. Rapid granular flows. Annu. Rev. Fluid Mech., 35, 267–293.
Hsu, T.J. and Jenkins, J.T. and Liu, P.L.F., 2003. On two-phase sediment transport: Dilute flow. J. Geophys. Res., Oceans 108(C3).
Jaeger, H.M., Nagel, S.R. and Behringer, R.P., 1996. Granular solids, liquids, and gases. Rev. Mod. Phys., 68(4).

Sediment transport in the Schelde-estuary: A comparison between measurements, transport formula and numerical models

Y. Plancke
Flanders Hydraulics Research, Antwerp, Belgium
Antwerp University, Antwerp, Belgium

G. Vos
Antea Group, Antwerp, Belgium
Flanders Hydraulics Research, Antwerp, Belgium

ABSTRACT

The Schelde-estuary serves different estuarine functions and therefore faces managers with multiple challenges: increasing tidal propagation vs. safety against flooding; sedimentation in the navigation channel vs. port accessibility; changing dynamics vs. ecology. Within the Flemish-Dutch Long Term Vision for the Schelde-estuary, a 4 year (2014–2017) research programme was defined, in which 8 topics will be dealt with (e.g. tidal penetration, risk for regime shift, sediment strategies, valuing ecology). Two fundamental tools are crucial in answering the different questions towards the future management of the estuary: expertise/system understanding and numerical models. Where the numerical models reproduce the hydrodynamics reasonably well, sediment transport and the resulting morphological changes is still a big challenge. In the past, the models have been extensively calibrated and validated for hydrodynamics, but due to lack of available measurement data, sediment transport was never really validated.

In 2014, a series of field measurement campaigns were organized to collect datasets for numerical model calibration and validation with regard to sediment transport and morphology. In total more than 10 locations were monitored over a full tidal cycle (13h), using both direct (Delft bottle, pump samples) and indirect techniques (optical and acoustical backscatter). Where measuring sediment transport remains very challenging, results have an uncertainty due to (a) measurements techniques (e.g. sensitivity of indirect techniques to sediment properties, bio-fouling), (b) measurement execution (e.g. errors made during direct sampling in sample collection), (c) field conditions (external factors e.g. ships) and (d) data post-processing (e.g. calculation of vertical profile based on discrete data).

Within the scope of several projects, these new datasets were used to validate numerical models. Within this paper the results of a validation was described for a 2D numerical model in Delft3D used to optimize the relocation strategy of dredged non-cohesive sediments in the Zeeschelde.

Sensitivity exercises have indicated the important influence of several numerical parameters. The comparison of field data and numerical model results show a rather promising agreement (i.e. differences of factor 2 to 3) for the Beneden-Zeeschelde. During certain moments of the tidal cycle sediment transport patterns are reproduced rather well, while at certain moments differences become larger. Where sediment transport is calculated with different formulas (e.g. Engelund-Hansen, Van Rijn) using flow velocities as an important parameter, minor differences in flow velocity are translated into large differences in sediment transport.

Finally a comparison with observed topo-bathymetric changes was made using the morphological module. It was found that important differences (both in patterns, intensities and between the 2 formula) were found. Therefore, it was decided that scenarios within the project to optimize the relocation strategy could not be done using the morphological model. Therefore different scenarios were investigated by making changes in the initial bathymetry (adding sediment to reproduce the relocation) and analyzing changes in flow patterns and sediment transport rates.

Influence of non-uniform flow conditions on riverbed stability: The case of smooth-to-rough transitions

D. Duma, S. Erpicum, P. Archambeau, M. Pirotton & B. Dewals
Research Group Hydraulics in Environmental and Civil Engineering (HECE), ArGEnCo Department, University of Liège (ULg), Liège, Belgium

ABSTRACT

Predicting the entrainment of bed material by water flow is of utmost practical relevance, particularly for the safe design of hydraulic structures and to ensure riverbank stability. The inception of sediment motion was studied extensively for a broad range of bed material properties and bed slopes; but most of these previous studies were based on the assumption of uniform flow conditions. In particular, many of them used a friction formula for evaluating the bed shear stress.

Recently, Hoan et al. (2011) and a few other researchers explored new avenues to assess the stability of riverbed material. They linked directly the sediment pick-up rate to the local flow and turbulence characteristics. Such approaches avoid the need for closure relations valid only under uniform flow conditions, such as most friction formula. However, they were validated only for a limited range of flow conditions and geometric setups (e.g., for sudden expansions).

In the present research, we aim to investigate whether such a new approach also holds in the case of a different geometric setup, namely for the case of smooth-to-rough transitions (Fig. 1). This configuration is very often observed downstream of hydraulic structures, at the transition between the concrete lined bottom and the natural riverbed.

We performed flume experiments considering two different configurations. We used a uniform bed material to define a reference configuration (C1) and we compared to a sudden transition from smooth to rough bottom (C2, Fig. 1). Two different flumes were used to assess possible scale effects and to extend the range of tested flow conditions. The bed material (diameter $d = 8$, 15 and 30 mm, relative density $\Delta = 0.5$, 0.7 and 1.65), the flume slope and the flow rate were varied. Using a Vectrino II ADVP (manufactured by Nortek) and a UVP probe (manufactured by METFLOW), flow velocity was measured at a high frequency (100 Hz) along different profiles in the near-field of the smooth-to-rough transition. The sediment pick-up rate was evaluated by counting the number of displaced particles.

Measurements of flow velocity and turbulent fluctuations were used to compute the non-dimensional

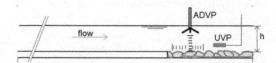

Figure 1. Tested geometric configuration for smooth-to-rough transition (side view).

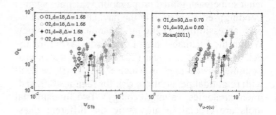

Figure 2. Observed bed mobility parameter Φ_E, as a function of Shields parameter Ψ_{S1b} (left) and the bed stability parameter $\Psi_{u-\sigma(u)}$ proposed by Hoan et al. (2011, right).

stability parameters $\Psi_{u-\sigma(u)}$ introduced by Hoan et al. (2011). The sediment pick-up rate was expressed in non-dimensional form through the bed mobility parameter Φ_E, also defined by Hoan et al. (2011). In Fig. 2a (respect. Fig. 2b), we display the observed bed mobility parameter as a function of the standard Shields parameter Ψ_{S1b} (respect. the bed stability parameter $\Psi_{u-\sigma(u)}$ of Hoan et al., 2011). Despite some scatter in the experimental results, they suggest a more significant correlation between the mobility parameter Φ_E and the new bed stability parameter $\Psi_{u-\sigma(u)}$; rather than between Φ_E and Shields parameter Ψ_{S1b}, which was evaluated assuming uniform flow conditions.

This study confirms the potential of this new approach based on local flow and turbulence characteristics. Additional experiments remain necessary to parametrize the relationships between the bed mobility parameter Φ_E and bed stability parameters, which were tested so far only in a narrow range of setups.

REFERENCE

Hoan, N. T., Stive, M., Booij, R., Hofland, B., Verhagen, H. J. (2011). Stone Stability in Nonuniform Flow. *Journal of Hydraulic Engineering.* 137(9): 884–893.

The probabilistic solution of dike breaching due to overtopping

Z. Alhasan, D. Duchan & J. Riha
Institute of Water Structures, Brno University of Technology, Czech Republic

ABSTRACT

This paper contains the results of the probabilistic solution of the breaching of a left bank dike of the River Dyje at a location adjacent to the village of Ladna near the town of Breclav in the Czech Republic.

The time course of a dike breach from the flood beginning until the total dike collapse is described by the following parameters:

– The time of the flood beginning: $t = 0$,
– The time t_o of the beginning of dike overtopping,
– The time t_e of the beginning of dike erosion,
– The time t_b of dike breach initiation,
– At time t_{max} the maximum breach discharge Q_{bmax} and maximum breach size are attained.

Among those instants, four typical phases of the dike breaching process were distinguished: flood wave arrival (no overtopping), resistance (overtopping – no erosion), breach initiation (erosion – no breaching) and breach formation (dike collapse).

A mathematical model describing the overtopping and breaching processes was proposed. The overtopping process is simulated using simple surface hydraulics equations. For modelling the dike erosion, a simple transport equation was used.

In this article, the dike breaching problem was divided into the process of dike overtopping followed by the process of erosion of the dike material. Therefore, the mathematical model was developed to consist of two parts (modules): a hydraulic module which describes the hydraulics of water flow during the dike overtopping process, and an erosion module that describes the progress of the erosion of the dike material.

The uncertainty in the input parameters of the dike breaching model was taken into account when selecting parameters for random sampling, in order to obtain a probabilistic solution for the problem. In this study the screening method was used to identify the non-influential input parameters. The most-used screening method in engineering is based on the so-called "One-At-a-Time" design, where each input is varied while keeping the others constant.

The assessment of the probability was related to the typical phases of the dike breaching process specified before. For the purpose of the probabilistic solution, a random sampling procedure was used where a set of simulations of dike breaching due to overtopping was generated with consideration that the value of each uncertain input parameter changes within an interval of values with a specific probability distribution. Using the Latin Hypercube Sampling (LHS) procedure, the sets of input parameters' values were randomly sampled and applied into the deterministic model to generate the set of output parameters. The probability P_i of each typical i-th phase of the dike breach was estimated by frequency analysis as follows:

$$P_i = \frac{\text{number of simulations realizing the phase}(i)}{\text{total number of simulations}}$$

The final results were presented as probabilities related to the annual occurrence of a given breaching phase. The assessment was carried out for three flood protection levels, namely against maximum discharges corresponding to the return periods $N = 10$, 20 and 50 years. The dike was covered by the plain grass which is a standard protection for the dike slope. The resulting probabilities are shown in Table 1.

Table 1. Probability values for the typical phases.

	No overtopping	Overtopping – no erosion	Erosion – no collapse	Collapse
$Z_c = h_s$ (Q_{10})	0.90595	0.00378	0.00003	0.09024
$Z_c = h_s$ (Q_{20})	0.95688	0.00198	0.00002	0.04112
$Z_c = h_s$ (Q_{50})	0.98543	0.00189	0.00001	0.01267

Field measurements and numerical modelling on local scour around a ferry slip structure

P. Penchev
Black-Sea Danube Coastal Research Association, Varna, Bulgaria

V. Bojkov
University of Architecture, Civil Engineering and Geodesy, Sofia, Bulgaria

V. Penchev
Black-Sea Danube Coastal Research Association, Varna, Bulgaria

ABSTRACT

This paper reviews a case study on a ferry slip near the town of Nikopol, Bulgaria, located at the Lower Danube River. This ferry slip is an earth and rock-fill hydraulic structure designed to operate as a terminal for passengers and cargo. The ferry slip with a complex spatial geometry was constructed back in 2006 on a very specific site along the Lower Danube. This eventually led to some stability problems, caused by the complex dynamics of the river flow, the contracted cross section, the helical flow along the structure, triggering erosion processes at the foot of the structure.

Structural changes were observed in the ferry slip in 2007, which required detailed hydrographic measurements including river flow velocities, bathymetry, sediments and geodetic works for the overall assessment of the causes that led to the problems with the structure. During the measurements in 2007 it was observed that there was a local scour about 3.0 m deep at the foot of the structure. After discovering the scour it was decided to put rock-fill mattresses to strengthen the river bed and stop the eroding process. Three rows of Reno mattresses with dimensions 6 × 2 × 0.3 m were set in place with the assistance of scuba divers.

Detailed hydrographic measurements were carried out in June 2015 to assess the current situation and to analyze the erosion processes in the river reach. It was discovered that the reinforced with rock-filled mattresses part of the river bed is stable and eventually the scour had seized, however it was found out that the erosion processes are continuing just downstream of this zone and a new scour hole evolved along the structure at its end.

The data collected during the hydrographic measurements allowed to establish a hydrodynamic numerical model for hindcast simulation of both hydraulic and morphological processes.

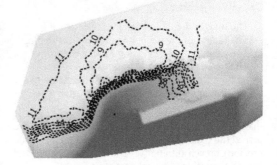

Figure 1. Local scour around the ferry slip in 2007.

MIKE 21 Flow Model numerical model for both hydrodynamic and morphological computations was used to simulate hydraulic and morphological processes in the investigated area. The results from hydrographic surveys were used to calibrate the numerical model, while archive data was used for the verification. The numerical simulations elaborated in this study showed good correspondence to the data collected during the field measurements in 2015. The erosion rate is computed to be approximately 1 cm/day during steady state simulation with an average discharge (Q_{avg}). The accumulation rate is also estimated to be in the same magnitude. The results from both numerical modelling and field measurements were further processed in GIS environment for the actual identification and calculation of the erosion/accumulation.

The results obtained provide a good ground for further use of the developed numerical model, in combination with the demonstrated GIS approach, for future forecasts of the dynamic of the river bed in this area.

Sustainable Hydraulics in the Era of Global Change – Erpicum et al. (Eds.)
© 2016 Taylor & Francis Group, London, ISBN 978-1-138-02977-4

Experimental and numerical study of scour downstream Toachi Dam

Luis G. Castillo
Hidr@m Group, Civil Engineering Department, Universidad Politécnica de Cartagena, Spain

Marco Castro
Centro de Investigaciones y Estudios en Recursos Hídricos (CIERHI), Escuela Politécnica Nacional, Ecuador

José M. Carrillo
Hidr@m Group, Civil Engineering Department, Universidad Politécnica de Cartagena, Spain

Daniel Hermosa, Ximena Hidalgo & Patricio Ortega
Centro de Investigaciones y Estudios en Recursos Hídricos (CIERHI), Escuela Politécnica Nacional, Ecuador

ABSTRACT

The study analyzes the expected changes in the Toachi River (Ecuador) as a result of the construction of the Toachi Dam (owned by CELEC EP Hidrotoapi). It is necessary to know the shape and dimensions of the scour generated downstream of the dam. Toachi is a concrete dam with a maximum height of 59 m to the foundations. The top level is located at an altitude of 973.00 meters above sea level. With normal maximum water level located at 970.00 m, the reservoir has a length of 1.30 km in the Sarapullo River and 3.20 km in the Toachi River. The dam has a free surface weir controlled by two radial gates. It consists in 2 channels located in centre of the dam that end in ski jump. The discharge is controlled by radial gates in order to ensure the accurate operation in the event when the gates are partially open. The spillway has been designed to spill the design flow (1213 m³/s). The scours downstream of the dam is studied with four complementary procedures: laboratory model with 1:50 Froude scale similitude, empirical formulae obtained in models and prototypes, semi-empirical methodology based on pressure fluctuations-erodibility index, and computational fluid dynamics (CFD) simulations.

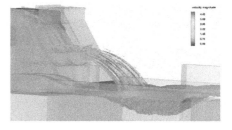

Figure 3. Numerical simulation of the scour downstream Toachi Dam.

Figure 1. Three-dimensional and physical model of the Toachi Dam.

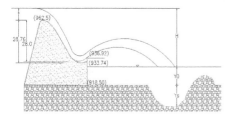

Figure 2. Scheme of scour in Toachi Dam.

REFERENCES

Castillo, L. G. and Carrillo, J. M., 2015. Characterization of the dynamic actions and scour estimation downstream of a dam. Proc. Int. Conf. Dam Protections against Overtopping and Accidental Leakage: 231–243. Madrid, Spain.

Castillo, L. G., Carrillo, J. M. and Blázquez, A., 2015. Plunge pool mean dynamic pressures: a temporal analysis in nappe flow case. Journal of Hydraulic Research 53(1): 101–118.

Escuela Politécnica Nacional, HIDROTOAPI E.P. 2013. *Estudio experimental en modelo hidráulico. Escala 1:50. Verificación experimental del diseño definitivo de la presa Toachi y obras complementarias. Informe Técnico Fase IV.* Quito, Ecuador.

Design of scour protections and structural reliability techniques

Tiago Fazeres-Ferradosa, Francisco Taveira-Pinto & Luciana das Neves
Faculty of Engineering, University of Porto, Portugal

Maria Teresa Reis
National Laboratory for Civil Engineering, Portugal

ABSTRACT

The renewable energy market is growing, around 2342 MW of offshore wind capacity were installed, during the first half of 2015.

Monopiles are commonly employed as part of the foundations for offshore wind turbines.

One of the major problems affecting the design of monopiles is the scour phenomena, which can affect the stability of the structure, namely reducing their ultimate capacity and the dynamic response (Prendergast et al., 2015), eventually leading to collapse.

In order to account for scour problems, protection systems are usually employed around the foundations, i.e. Arklow Bank OWF and Scroby Sands OWF (Whitehouse et al., 2011).

Their design is mainly of semi empiric nature (De Vos et al., 2011) and probabilistic approaches haven't yet reached a mature state of development, for cases of waves and currents combined.

Reliability-based methodologies imply the calculation of failure probabilities, which enable to provide a measure of risk instead of the typical safety factor concept. Reliability techniques have been mainly developed as structural design methodologies, typically applied in structural elements, such as piles, beams and other elements of buildings' design. Some works have extended the reliability methodology to scour phenomena, in fluvial conditions.

However, the lack of methodologies for the reliability-based design of scour protections in offshore environment is evident.

The present work aims to provide the preliminary results of a reliability based methodology, to design rip-rap systems, around monopile foundations in marine environment. This procedure was based on Monte Carlo Method coupled with the Latin Hypercube Algorithm, to minimise the number of simulations needed for the probabilities' convergence.

The results provided in this investigation allowed to establish a reliability-based methodology to quantify the risk associated to the global safety factor.

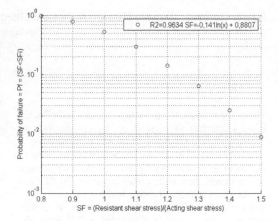

Figure 1. Safety Factors *vs.* Probability of Failure for the protection.

The results showed that despite the decreasing probability of failure with the increasing safety factor, a false notion of security might be induced.

For example, with a safety factor of 1.1, for which the top layer of the protection presents a shear stress resistance 10% higher than the one generated by the sea state conditions, the protection still enters into failure 29% of the times.

REFERENCES

De Vos L., Rouck J., Troch P., Frigaard P. 2011. Empirical design of scour protections around monopile foundations. Part 1: Static approach, Coastal Engineering, vol. 58, pp. 540–553.

Prendergast, L., Gavin, K., Doherty, P. 2015. *An investigation into the effect of scour on the natural frequency of an offshore wind turbine.* Ocean Engineering, Vol. 101, pp. 1–11. DOI: 10.1016/j.oceaneng.2015.04.017.

Whitehouse, R., Harris, J., Sutherland, J., Rees, J. 2011. *The nature of scour development and scour protection at offshore windfarm foundations.* Marine Pollution Bulletin, Vol. 62, pp. 73–88, DOI: 10.1016/j.marpolbul.2010.09.007.

Failure of fluvial dikes: How does the flow in the main channel influence the breach development?

I. Rifai
ArGEnCo Department, Research Group Hydraulics in Environmental and Civil Engineering (HECE), University of Liège, Liège, Belgium
EDF R&D, National Laboratory for Hydraulics and Environment, Chatou, France
Saint Venant Laboratory for Hydraulics, Chatou, France

S. Erpicum & P. Archambeau
ArGEnCo Department, Research Group Hydraulics in Environmental and Civil Engineering (HECE), University of Liège, Liège, Belgium

D. Violeau
EDF R&D, National Laboratory for Hydraulics and Environment, Chatou, France
Saint Venant Laboratory for Hydraulics, Chatou, France

M. Pirotton & B. Dewals
ArGEnCo Department, Research Group Hydraulics in Environmental and Civil Engineering (HECE), University of Liège, Liège, Belgium

K. El kadi Abderrezzak
Saint Venant Laboratory for Hydraulics, Chatou, France

ABSTRACT

Fluvial dikes are commonly constructed for flow channelization and as flood defences; but in case of failure they lead to significant casualties and damages. Statistics show that overtopping is the first cause of failure (Fry et al. 2012; Foster et al. 2000). Many experimental studies on dike failure were conducted, but most of them focused on normal configurations (i.e. where the flow is perpendicular to the dike axis, such as in a dam break configuration), without accounting for the influence of the flow parallel to the dike in the main channel. The aim of the present work is to improve the current understanding of physical processes underpinning gradual breaching of fluvial dikes by overtopping, taking into account the flow parallel to the dike axis.

One challenging aspect of laboratory experiments of dike failure is the continuous monitoring of the breach evolution. We have developed a non-intrusive and distributed measurement technique based on laser profilometry (Figure 1). The approach offers a high degree of flexibility, since neither the location of the camera nor that of the laser need to be known *a priori* (Rifai et al. 2016).

Results of the experimental tests unveil breach evolution mechanisms that strongly differ from those occurring in a frontal

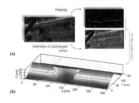

Figure 1. Main steps of the continuous 3D reconstruction of the breach geometry by the laser profilometry: (a) images of the experimental dike failure and processing, (b) final reconstruction of the dike in real-world coordinates.

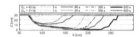

Figure 2. Evolution of the breach centerline section for two different channel discharges Q_{in}.

dike failure configuration. The breach expands almost only towards downstream of the main channel. This expansion is caused to a great extent by slope failure, whereas the limited widening of the breach towards upstream is controlled by the acceleration of the flow in the main channel and the resulting erosion of the dike material. This yields to a profiled shape of the upstream part of the dike. The flow conditions in the main channel significantly affect the duration of the breaching process and the evolution of the width-to-depth ratio of the breach (Figure 2).

These results confirm that the present knowledge on dam breaching cannot be transposed to fluvial dike cases, i.e. fluvial dike overtopping involves physical processes specific to the fluvial configuration.

The correlation of the breach evolution dynamics with flow conditions and channel configuration will allow reliable transposition of the experimental findings to real-world fluvial dikes.

REFERENCES

Foster, M., Fell, R. and Spannagle, M., 2000. The statistics of embankment dam failures and accidents. Canadian Geotechnical Journal, 37(5), pp. 1000–1024.

Fry, J., Vogel, A., Royet, P. and Courivaud, J., 2012. Dam failures by erosion: lessons from ERINOH data bases Key words., pp. 273–280.

Rifai, I., Erpicum, S., Archambeau, P., Violeau, D., Pirotton, M., El kadi Abderrezzak, K. and Dewals, B., 2016. Monitoring topography of laboratory fluvial dike models subjected to breaching based on a laser profilometry technique. International Symposium on River Sedimentation, Stuttgart.

Continuous grid monitoring to support sediment management techniques

T. Van Hoestenberghe & R. Vanthillo
Fluves, Ghent, Belgium

P. Heidinger & J. Dornstädter
GTC Kappelmeyer, Germany

N. Dezillie & N. Van Ransbeeck
VMM, Brussels, Belgium

ABSTRACT

Fluves monitors since November 2015 continuously a sediment trap with dimensions 200 × 20 meters managed by the Flemish Environmental Agency (VMM) in Belgium. The continuous follow-up of sedimentation of the trap provides insights on temporal and spatial evolution of trapping efficiency. VMM will use the insights to optimize operational dredging decisions for the trap and for optimizing the design of future traps. The installed measuring system is based on distributed temperature sensing with a fiber optic cable of more than 2 km.

The trap is installed in a flood control area (FCA) on the river 'Moenebroekbeek', an upstream branch of the Dender river, a tributary of the Scheldt river. Average river discharge is 500 l/s. The sediment type that enters the FCA is fine loam. The estimated annual sediment transport towards the trap is 1500 ton.

The sediment height is measured throughout the pool by combining direct and indirect measurements. At 20 locations, evenly distributed in the pool, the soil/water interface is measured directly by winding the fiber optic cable around vertical poles (Figure 1). In between the poles, the cable is installed horizontally on top of the base level of the sediment trap and sediment height is measured indirectly.

The DTS unit measures the thermal response of the heating for all measurement points along the cable, one value every 25 cm.

When the cable is surrounded by water, the thermal response will be high. When the cable is buried

Figure 2. Sedimentation height in centimeters between November 2015 and begin February 2016 at the 20 direct measurement locations. Red and brown zones indicate the locations with respectively less and more sedimentation than the average sedimentation rate (i.e. 5 cm).

in sediments, the response will be low. Because the installation layout is known, the results can be transformed from fiber meter distance to height of the poles. For the vertical poles, the threshold from low to high thermal responses marks the upper boundary of sedimentation. Sediment height is directly measured with an accuracy of 1 cm at the 20 poles. The results of sedimentation height in centimeters between November 2015 and begin February 2016 are illustrated in Figure 2. The time series of sedimentation show areas with slower sedimentation, but also areas where first sedimentation and erosion succeeded.

The technique shows great potential for detailed spatial and temporal observation of sedimentation processes in large areas.

Figure 1. Sediment trap after installation of the monitoring system with 20 vertical poles throughout the trap.

River flow analysis with adjoints – An efficient, universal methodology to quantify spatial interactions and sensitivities

U.H. Merkel
UHM River Engineering, Karlsruhe, Germany

J. Riehme & U. Naumann
STCE, RWTH Aachen University, Germany

ABSTRACT

Sensitivities are the key to uncertainties approximation and calibration or optimization strategies. But they are very expensive to calculate for many real world problems of the inverse base type.

This article describes a new method for open channel flow software which solves inverse flow problems. Instead of forward questions like

"Put water in here, how will it be distributed within our project area?"

we solve *INVERSE* questions of the type

"We want specific flow related project result, where and how do we have to modify our river?".

See figure 1 for a simple example comparison:

– Fig. 1 (left) shows the traditional forward project: Few input parameters that influence many target parameters.
– Fig. 1 (middle) shows a typical inverse project: Many input parameters have influence on one target parameter.
– Fig. 1 (right) shows the solution of the inverse problem through backward interpretation of forward calculations.

If the hydraulic problem is based on a very large number of influence factors, then the adjoint model of the TELEMAC-SUITE is the most efficient solver: All sensitivities/gradients are computed by one run of the adjoint model. Which is a speed up of more then × 1000 for many every day problems.

A wide range of new applications is now possible on basis of well validated source code: Automatic optimization and calibration of flow relevant shapes, data assimilation, high resolution sensitivity analysis or deciphering of superimposing flow effects.

The full paper shows several examples for a river engineer's daily business and method explanations.

Additional information is available under: www.uwe-merkel.com/TELEMAC-AD.

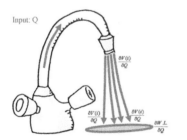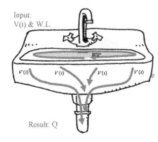

Figure 1. Left: Common forward problem: Few parameters influence many, dependencies easy to calculate. Middle: Inverse problem. Few parameters depend on many, sometimes millions of geometry points. Right: The adjoint method solves until now unsolvable inverse problems by interpreting the forward problem backward and producing up to millions of independent sensitivities in one work step.

Comparative use of FVM and integral approach for computation of water flow in a coiled pipe and a surge tank

Muris Torlak & Berin Šeta[1]
University of Sarajevo – Mechanical Engineering Faculty, Sarajevo, Bosnia-Herzegovina
[1]*TENSOR Technology Doo, Sarajevo, Bosnia-Herzegovina*

Adis Bubalo
Elektroprivreda BiH – HE na Neretvi, Jablanica, Bosnia-Herzegovina

ABSTRACT

Results of combined 3D *computational fluid dynamics* (CFD) analysis and 1D integral analysis of water flow in hydraulic systems are presented. CFD analysis is based on solution of Navier-Stokes equations for incompressible flows using a *finite-volume method* (FVM), which requires discretization of space, time and governing equations, leading thus to a relatively large computational effort, but delivering detailed information on various flow properties. Anticipating faster computations, tailored methods for integral 1D analysis of two different problems considered in this paper are also applied. The examples presented include (*i*) FVM calculation of pressure drops and corresponding friction factors in a helically coiled pipe at different flow regimes, which are compared with the results of relations from literature implemented in the integral model for curved pipes (Fig. 1), as well as (*ii*) the use of another integral model and FVM for calculation of water elevation in a surge tank of a hydropower plant, for which measured data are available (Fig. 2).

The examples show that CFD calculation provides detailed information on flow properties relying upon the detailed physical modeling, which requires however significantly longer computing times. While for analysis of laminar pipe flow relatively coarse discretization may be used without significant impact on the accuracy, reliable prediction of turbulent pipe flow and proper capturing of the free surface flow in the surge tank are manageable, but require detailed discretization.

On the other hand, both integral models (for pipe systems and for surge tanks) implemented in two different in-house computer programs deliver satisfactory accuracy, approaching the CFD results at notably shorter computing times. Their accuracy depends on the friction factor and minor-loss coefficient, since detailed physical effects are not resolved by the models. Having known these values from empirical data or CFD simulations of reduced models to shorten necessary computing times, compromise solution can be achieved.

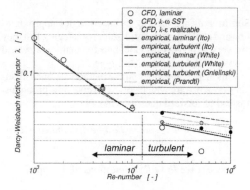

Figure 1. Friction factor in a helical coil pipe: comparison between CFD-calculated and empirical values.

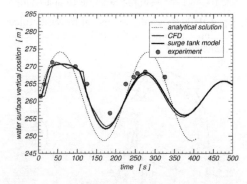

Figure 2. Time history of the water level in the surge tank for the flow rate of 72 m^3/s with the initial water level 44.4 m.

REFERENCES

CD-adapco. 2014. *User Guide, STAR-CCM+, Version 9.04*. www.cd-adapco.com.

Ferziger, J.H. & Perić, M. 2013. *Computational Methods for Fluid Dynamics*. Springer.

Seepage characteristics in embankments subject to variable water storages on both sides

M. Calamak
Department of Civil Engineering, TED University, Ankara, Turkey

A.M. Yanmaz
Department of Civil Engineering, Middle East Technical University, Ankara, Turkey

ABSTRACT

Long embankments are constructed on plains to support highways and railways. The common modes of failure in an embankment are attributed to overtopping of the crest by a flood wave, occurrence of excessive seepage through the embankment during a flood, failure of the slopes, and the reduction of the crest level due to excessive settlement (Morris et al. 2007). This study investigates the seepage characteristics of embankments under flooding conditions using SEEP/W®. A realistic embankment is defined in a hypothetical application and modeled for seepage under transient boundary conditions at its both sides.

The results of the analyses showed that the seepage wetting line propagated to the center of the embankment during the rising stage of the flood from both sides. During this period, the pore water pressures at the upstream and downstream sides and at the foundation of the embankment increased. Then, during the falling stage, both the pore water pressures and the seepage fluxes started to decrease. The wetting line of the upstream side never reached to the downstream side. Similarly, the wetting line propagating from downstream side was never reached to the upstream side. The central part of the embankment was kept dry during the flood. Therefore, it can be concluded that no protective measures are needed to decrease the seepage for such embankments subject to variable water levels on both sides.

According to the results, the critical hydraulic gradient was only exceeded during the rising stage of the flood in the variable fill layer close to the upstream face of the embankment. However, the duration of this was inspected to be relatively short and for this instance the flow direction was from upstream face to the center of the embankment. If any soil particle is moved from its place, it will eventually stop in the vicinity of the embankment center. Therefore, it can be said that piping is not expected in the embankment subject to variable water levels on both sides and there is no need to implement any facility against it.

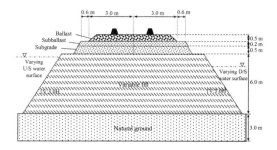

Figure 1. The cross-section of the railway embankment modelled in the study (Not to scale).

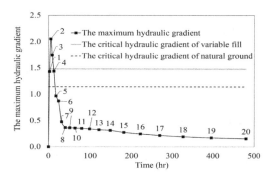

Figure 2. The change of the maximum gradient in the embankment with respect to time.

In view of the embankment safety, slope stability analyses are also conducted. The minimum factor of safety values are approximately 1.70 and 1.65 for the upstream and downstream slopes, respectively. The railway embankment is found to be safe against slope failure.

REFERENCE

Morris, M., Dyer, M., and Smith, P. (2007). *Management of Flood Embankments: A good practice review*. London.

Characterization of nappe vibration on a weir

M. Lodomez, M. Pirotton, B. Dewals, P. Archambeau & S. Erpicum
University of Liège, Argenco Department, Liège, Belgium

ABSTRACT

As a key component of dams safety, weirs are commonly used as spillways to release flows from a reservoir. The free falling jet on the downstream side of the weir, called nappe, can display a variety of behaviors. Under relatively low-head discharges, the behavior of nappe oscillation, otherwise known as nappe vibration, can occur. This phenomenon classified by Naudascher & Rockwell (1994) as an instability induced excitation, has been also observed on hydraulic structures with a free overfall, such as crest gates and fountains. Early identified as a undesirable and potentially dangerous, the most recognizable characteristics of this phenomenon are the horizontal banding on the nappe and the extreme acoustic energy resulting in low frequency noise that can be heard in close proximity of the structure.

Twenty years after the early works, the phenomenon is far not well understood and poorly controlled. Recently new investigations have been undertaken at the Utah Water Research Laboratory (Utah State University) (Anderson 2014; Crookston et al. 2014). The recent observations with some conflicting results from previous analysis confirm the need for systematic data measurement on large-scale models to come up with generic scientific conclusions as well as a deeper systematic analysis of scale effects.

In this context, a detailed investigation of nappe vibration is being conducted at the Engineering Hydraulics laboratory of the University de Liège on a prototype-scale linear weir. As illustrated in Figure 1, the physical model is an elevated box divided in two identical weirs with a 3.46-m long crest and a 3.04-m high chute. The air cavity behind the nappe can be confined or vented to the atmosphere.

To identify and quantitatively characterize the nappe oscillation, sound and image analysis procedures have been developed and applied to confined

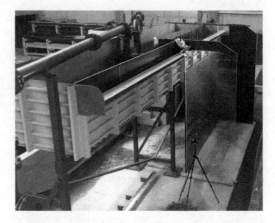

Figure 1. Experimental model.

and vented models. This paper provides an overview of the preliminary results regarding the characterization of the nappe vibrations. In particular, the flow range for which nappe vibrations appear, the noise generated by the phenomenon and the frequency vibrations have been studied and quantified.

REFERENCES

Anderson, A.A. 2014. *Master thesis: causes and countermesures for nappe oscillation*. Utah State University.
Crookston, B.M., Anderson, A., Shearin-Feimster, L. & Tullis, B. P. 2014. Mitigation investigation of flow-induced vibrations at a rehabilitated spillway. In *ISHS 2014 – Hydraulic Structures and Society – Engineering Challenges and Extremes: Proceedings of the 5th IAHR International Symposium on Hydraulic Structures*.
Naudascher, E. & Rockwell, D. 1994. *Flow-induced vibrations: an engineering guide*, Balkema.

Shock wave patterns in supercritical junction manholes with inlet bottom offsets

G. Crispino & C. Gisonni
Department of Civil Engineering, Design, Building and Environment, Second University of Naples, Aversa, Italy

M. Pfister
Hydraulic Constructions Laboratory, École Polytechnique Fédérale de Lausanne, Lausanne, Switzerland
Civil Engineering Department, Haute Ecole d'Ingénierie et d'Architecture, Fribourg, Switzerland

ABSTRACT

In the last decades, severe floods have frequently hit urban areas. The catastrophic effects of such events are even multiplied if sewers are affected by hydraulic weakness and deficiency. Many times, drainage system failures are generated by poor flow conditions in sewer manholes, which are not always adequately designed. Among various sewer manholes, junction manholes represent basic elements for sewers because they enable two or more upstream branches to merge into a single downstream branch. The flow junction is a complex structure, especially for supercritical flow conditions. In such cases, the theoretical modelling is made complicate by the occurrence of particular phenomena (i.e. shock waves, breakdowns, choking etc.). The semi-empirical approaches based on physical model investigation are thus more indicated.

First experimental studies were conducted to study the hydraulics of 45° and 90° junction manholes under both sub- and supercritical conditions (Del Giudice & Hager 2001, Gisonni & Hager 2002). Several tests were performed focusing on junctions with a fixed branches diameter and for an even invert, i.e. all conduits bottoms and tops were located at the same elevations. Conversely, practical sewer applications resort commonly to the junction arrangement with aligned conduit tops, mainly to avoid backwater effects. When upstream and downstream branch diameters differ, bottom offsets at manhole inlets subsist.

The presence of bottom offsets may change the hydraulic features of junction manholes. For this reason, an extensive set of experimental tests was carried out at the Laboratory of Hydraulic Constructions (LCH) of École Polytechnique Fédérale de Lausanne (EPFL). The model (Fig. 1) reproduced the afore-mentioned junction manhole arrangement with aligned branch tops and variable branch diameters. It was composed of a straight and a lateral branch, merging under 45° in a junction chamber, from which a

Figure 1. Photograph of the junction manhole model.

downstream branch came out. The experimental investigation focused on supercritical flows entering the junction manhole.

The experimental data showed that the diagram of state defined by Del Giudice & Hager (2001) for 45° junction manholes need to be adapted to keep into account the effect of the inlet offsets. Shock wave B (Schwalt 1996) resulted to be the governing feature in the junction design. In fact, during certain test-runs, wave B provoked relevant water overflows and its impact on the manhole endwall produced massive spray development, interruption of the air entrainment along with shocked outflows.

REFERENCES

Del Giudice, G. & Hager, W.H. 2001. Supercritical flow in 45° junction manhole. *Journal of Irrigation and Drainage Engineering* 127(2): 100–108.

Gisonni, C. & Hager, W.H. 2002b. Supercritical flow in the 90° junction manhole. *Urban Water* 4: 363–372.

Schwalt, M. 1996. Vereinigung schiessender Abflüsse. *Gwf Wasser-Abwasser*, 137(6): 326–330 (in German).

Investigation of the hydrodynamic pressures on lock gates during earthquakes

L. Buldgen & P. Rigo
ANAST, University of Liège, Liège, Belgium

H. Le Sourne
GeM Institute, UMR CNRS 6183, ICAM Nantes, France

ABSTRACT

In seismic areas such Central America, the Middle East or the west coast of Latin America for example, the design of lock gates has to be done by accounting for the total hydrodynamic pressure that may be generated during an earthquake. This latter is known to be the sum of three contributions:

1) The convective pressure that is associated to the sloshing of the water contained in the lock chamber.
2) The rigid pressure that is generated by the motion of the gate as if it were a perfectly rigid body simply following the ground motions.
3) The flexible pressure that is due to the own vibrations of the gate as a deformable structure.

The two first contributions are already well known and have been extensively studied in the literature. They may be easily calculated by analytical formulae that are also available in international standard. The third contribution is however more difficult to evaluate because it depends on the fluid-structure interaction. Finding an analytical solution to this coupled problem is not straightforward. For this reason, finite element simulations are often needed to correctly capture the fluid-structure interaction. Unfortunately, these calculations are time demanding because both the solid and fluid domains have to modeled. Doing so is therefore not particularly suited for the early design of a lock gate, which obliged engineers to use other simplified approaches (such as the added-mass methods) that are not always relevant.

In order to circumvent these difficulties, this paper presents a new meshless method to evaluate the seismic pressure induced on a lock gate. In this analytical approach, the fluid is supposed to be irrotational, incompressible and inviscid. The Laplace equation is then used with appropriate boundary conditions that accounts for the flexibility of the gate, which leads to a coupled formulation that is solved in two steps:

1) A modal analysis of the gate is performed in order to get the mode shapes.
2) The dynamic analysis of the gate is performed by deriving a weak form of the equilibrium equation.

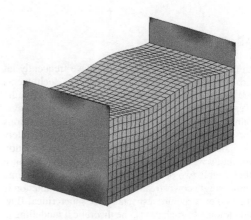

Figure 1. Water pressure in a lock chamber during a seism.

Based on the Galerkin method, a matrix formulation is obtained and the Newmark integration scheme is used for the evaluation of the pressure during the earthquake.

As a matter of validation, the solutions provided by this analytical approach are compared to the results of non-linear dynamic finite element simulations performed for a realistic lock gate (Figure 1). In these numerical simulations, the water is modeled as an elastic medium and the interaction with the gate is captured by imposing contact conditions between the gate and the fluid.

REFERENCES

Buldgen, L. 2015. A simplified analytical method to evaluate the seismic pressure on plane lock gates. *Engineering Structures* 100: 522–534.

Housner, G.W. 1957. Dynamic pressures on accelerated fluid containers. *Bulletin of Seismological Society of America* 47: 15–37.

Kim, J.K., Koh, H.M. & Kwahk, I.J. 1996. Dynamic response of rectangular flexible fluid containers. *Journal of Engineering Mechanics* 122: 807–817.

Stilling basin design for inlet sluice with vertical drop structure: Scale model results vs. literature formulae

J. Vercruysse & K. Verelst
Flanders Hydraulics Research, Antwerp, Belgium

T. De Mulder
Hydraulics Laboratory, Department of Civil Engineering, Ghent University, Ghent, Belgium

ABSTRACT

Within the framework of the Updated Sigmaplan, Flood control Areas (FCA) with a Controlled Reduced Tide (CRT) are set up in several polders along the tidal section of the river Scheldt and its tributaries. The reduced tide is introduced by means of simple inlet and outlet sluices located in the levee between the river and the polder. In recent designs, the inlet sluice is placed on top of the outlet sluice, a principle drawing is given in figure 1.

Flanders Hydraulics Research performed a scale model based review of different desktop designs of this type of combined inlet and outlet structures. Different types of sluice geometries – with a straight stilling basin, a locally deepened stilling basin and a stilling basin with baffle blocks – were studied. For the conjugate water depth and the drop length a comparison is carried out between scale model results and formulae from Rand (1955) and Chanson (2002). For the end of the hydraulic jump a comparison is carried out between scale model results and formulae from Rand (1955), Silvester (1964) and Hager et al. (1990).

For a straight stilling basin and a tailwater depth equal to the conjugate water depth, the comparison with literature formula shows that the conjugate water depth, the drop length and the hydraulic jump length can be predicted rather well using the available literature formulae. These formulae are also valid for a well-designed locally deepened stilling basin with an end sill of a limited height. For a stilling basin with baffle blocks designed according the design rules for a USBR type III stilling basin (Peterka, A.J. 1984), the conjugate water depth can be reduced with a factor 0.85.

The combined inlet and outlet structures for the FCA-CRT's will have to deal with a broad combination of upstream and downstream water levels. The increase of the tailwater depth results into a decrease of the angle of the falling nappe and a flow pattern varying from a free hydraulic jump, over a drowned hydraulic jump, to a free surface jet, to culvert flow. Especially for the geometry with a locally deepened stilling basin or baffle blocks, care should be taken that the falling nappe touches the bottom within the locally deepened stilling basin or upstream of the baffle blocks, in case of higher tailwater depths. Otherwise, these situations could lead to an increase in near bottom velocity downstream of the construction and degradation of the downstream bottom protection.

For higher tailwater depths than the conjugate water depth, no suitable literature formulae were found by the authors and physical (or numerical) modelling is recommended for designing the combined inlet and outlet structure.

REFERENCES

Chanson, H. (2002). The hydraulics of stepped chutes and spillways. Swets & Zeitlinger: Lisse. ISBN 90-5809-352-2.
Hager, W.H.; Bremen, R.; Kawagoshi, N. (1990). Classical hydraulic jump: length of roller. J. Hydraul. Res. 28(5): 591–608. doi:10.1080/00221689009499048
Peterka, A.J. (1984). Hydraulic design of stilling basins and energy dissipators. Engineering monograph (Washington), 25. U.S. Dept. of the Interior, Bureau of Reclamation: Washington.
Rand, W. (1955). Flow geometry at straight drop spillways. Proc. Am. Soc. Civ. Eng. 81: 1–13.
Silvester, R. (1964). Theory an dExepriment on the Hydraulic jump. Proc. Am. Soc. Civ. Eng. J. Hydraul. Div. vol. 90, n0 HY1 23–55.

Figure 1. Principle drawing of a combined inlet and outlet structure.

Supercritical flow around an emerged obstacle: Hydraulic jump or wall-jet-like bow-wave?

G. Vouaillat, N. Rivière, G. Launay & E. Mignot
Université de Lyon, INSA de Lyon, Laboratory of Fluid Mechanics and Acoustics, France

ABSTRACT

When an impervious, emerged, bluff obstacle is placed within a supercritical flow, the flow is forced to skirt the obstacle. The workaround can nevertheless take two distinctive forms.

A first one is the detached hydraulic jump, similar to the detached shock waves that forms upstream of bluff bodies in supersonic flows. Indeed, as the upstream flow is supercritical, the disturbances created by the obstacle cannot go back up the flow to promote a streamline curvature. The only solution for such a flow deflection is to form a subcritical zone upstream from the obstacle, through a detached hydraulic jump. The latter and its interaction with the horseshoe vortex that forms at the obstacle toe was already characterized (Defina and Susin, 2006; Mignot et Rivière, 2010; Rivière et al., 2012).

A second form of workaround is the so-called "wall-jet-like bow-wave" (Figure 1), which is in fact more commonly encountered in the field around bridge piers within rivers in supercritical regime. In this case, the workaround is directed upwards instead of laterally. On the upstream face of the obstacle, the water height is higher than with a hydraulic jump, as the energy dissipation is smaller.

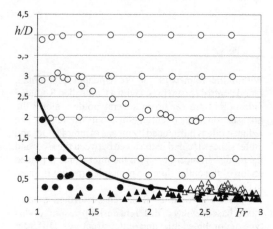

Figure 2. Flow forms observed depending on h/D and Fr: detached hydraulic jump (closed symbols) and wall-jet-like bow-wave (open symbols). Experiments using facility 1 sketched by triangles and using facility 2 by circles.

The present work, through laboratory experiments, details the two forms – jump and bow-wave. Then, using appropriate dimensionless numbers, it establishes the conditions of appearance of one form or another, which depend on the flow depth to obstacle with ratio h/D and on the upstream Froude number (Figure 2). Results show that more streamlined obstacles favor the appearance of a wall-jet.

REFERENCES

Defina A. and Susin F. M. 2006. Multiple states in open channel flow, in vorticity and turbulence effects in fluid structures interactions. Advances in Fluid Mechanics, edited by M. Brocchini and F. Trivellato _WIT Press, Southampton, UK, 2006 pp. 105–130.

Mignot,E. N. Rivière, 2010. "Bow-wave like hydraulic jump and horseshoe vortex around an emerged obstacle in a supercritical flow", Phys. Fluids, 22, 117105.

Rivière N., Laïly G., Mignot E., Doppler D. 2012. "Supercritical flow around and beneath a fixed obstacle", Proc. 2nd IAHR Europe Congress, Munich, Germany.

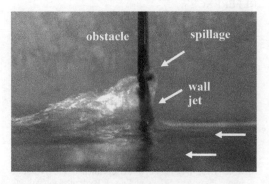

Figure 1. Example of wall jet like bow wave.

Monsin movable dam in Belgium: A case study

C. Swartenbroekx, C. Savary & D. Bousmar
Hydraulic Research Laboratory, Service Public de Wallonie, Châtelet, Belgium

ABSTRACT

Monsin movable dam is located on river Meuse, downstream of the city of Liège, in Belgium. It allows the water level in Albert canal to be maintained. It is thus a key dam for navigation between Liège, Anvers and Maastricht. It also regulates the water flow during high stage periods. The Walloon administration is considering its renovation since the present dam, built in 1930, is aging and adequate safety of the structure can no longer be guaranteed.

It is made of 6 27 m-wide sluices. Each opening is closed by a vertical lift valve on top of which is sat a flap gate. The whole structure is 5.35 m-high. Depending on the coming discharge (up to 2500 m³/s), the water can run as an overflow, an underflow or a mixed flow (Fig. 1). The ratio between the flap gate height (2 m instead of 1.96 m) and the lift valve height (3.35 m instead of 3.39 m) will be slightly modified to allow for more overflow, which is easier to regulate. Due to this ratio modification, several hydraulic behaviours had to be checked.

A 1:15 scale physical model of a slice of this dam was built in a 1 m-wide flume, according to Froude similarities (Fig. 2). It was used to determine (1) the efficiency of the aeration of the overflowing nappe; (2) the relationship between the discharge and the valve position; (3) the hydrodynamic pressure distribution on the valves; and (4) the energy dissipation on the downstream concrete apron.

The main conclusions are here summarized.

(1) The aeration can be guaranteed with a lateral spoiler and several breakers.
(2) The limit between overflow and mixed flow occurs for about $Q_p = 820$ m³/s while an underflow starts for about $Q_p = 2135$ m³/s. These limits correspond to less than 4% of time and about 0.02% of time, respectively.
(3) During an overflow, the hydrodynamic pressure is decreasing on the flap gate and the top of the lift valve in comparison with the hydrostatic profile. During an underflow, a gap between the hydrostatic profile and the hydrodynamic profiles is appearing at the valve toe.
(4) The present apron and dissipation structures are efficient enough to dissipate this energy, impeding excessive erosion of the mobile bed.

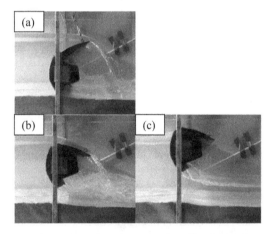

Figure 1. Lateral view of vertical lift valve and flap gate at (a) overflow, (b) mixed flow, and (c) underflow (physical model).

Figure 2. Side view of the physical model during a mixed flow. From left to right: the upstream water level, the valve, the flow over the downstream apron and the pit with mobile bed.

Numerical modelling of contracted sharp crested weirs

A. Duru
İller Bank, Ankara, Turkey

A.B. Altan-Sakarya
Middle East Technical University, Ankara, Turkey

M.A. Kokpinar
State Hydraulic Works, Artvin, Turkey

ABSTRACT

Sharp crested weirs are flow measurement structures which are frequently used for discharge measurements in channels and laboratories (Bos 1989). In this study, numerical modelling technique was used in the solution of contracted sharp crested weirs. The numerical simulation of the contracted sharp crested weir geometry studied by Aydın et al. 2011 was used. The basic flow equations are solved using the finite-volume method of the commercially available software Flow-3D. The volume of fluid method is used to compute the free surface of the flow that interacts with the discharge measurement structures.

To obtain the optimum simulation model conditions, the effects of mesh cell size, turbulence model, and upstream channel length were studied separately on a particular case with a defined upstream water depth. Then, after deciding the optimum model parameters with the specified physical definitions, the model was solved for the other water levels. The summary of the simulations is given in Table 1. The comparison of numerical and experimental results is given in Figure 1.

Optimization study proved the importance of selection of optimum mesh size in numerical simulations. Although minimizing the mesh cell size gives better results, it increases the computation time dramatically.

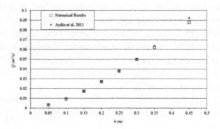

Figure 1. Comparison of the results.

It was deduced that mesh sizes of 4 mm was adequate to obtain accurate simulation results.

RNG and LES turbulence models were tested to find the effect of turbulence model on the results of numerical simulations. Both models gave similar results be due to the fact that turbulence effect in the subcritical approach flow of the rectangular sharp-crested weirs is of little to no importance. RNG model was used in the rest of the simulations.

The variable upstream channel lengths were studied in the scope of the optimization study. Although similar results were obtained for all lengths, upstream channel length of 2.5 m was accepted as sufficient to obtain accurate results.

It was shown that the formulation given by Aydin et al. 2011 related to discharge coefficients for contracted sharp crested weirs valid both within and up to a certain limit for outside the experimental range.

Eventually, it can be stated that CFD is a powerful tool in the solutions of hydraulic problems related to flow measurement structures.

Table 1. Summary of model test parameters affecting numerical simulation results.

Model parameter tested	Notes
Mesh cell size	4, 6 and 10 mm uniform mesh sizes are tested.
Turbulence model	RNG and LES turbulence models are used within and outside the experimental data ranges.
Upstream channel length	1, 1.5, and 2.5 m upstream channel lengths are tested.

REFERENCES

Aydın, İ. Altan-Sakarya, A. B. & Şisman, Ç. 2011. Discharge formula for rectangular sharp-crested weirs. Journal of Flow Measurement and Instrumentation 22: 144–151.

Bos, M. G. 1989. Discharge measurement structures. Wageningen: International Institute for Land Reclamation and Improvement.

Flow Inc., Flow 3D User Manual v10.

Field measurements at the new lock of Lanaye (Belgium) before the opening to navigation

C. Savary, D. Bousmar, C. Swartenbroekx & G. Zorzan
Hydraulic Research Laboratory, Service Public de Wallonie, Châtelet, Belgium

T. Auguste
Engineering Structural Expertise, Service Public de Wallonie, Liège, Belgium

ABSTRACT

In the frame of the 18th priority project of the Trans European Network of Transport (TEN-T) concerning the corridor Rhine-Main-Danube, a new navigation lock was inaugurated in 2015 in Lanaye on the borderline between Belgium and The Netherlands. The new lock is 225 m long and 25 m wide and will allow vessels up to 9000 tons (CEMT Class VIb) to cross a difference in height of 13.7 m between the Albert canal and the Lanaye canal.

The filling and emptying system is composed of longitudinal culverts with side ports, the filling and emptying discharge is regulated with butterfly valves of 3.5 m diameter. During the trial period, it was noticed that the structures were submitted to vibrations during the lockage and that cavitation occurred around the butterfly valves.

Before opening the lock to navigation, different measurements were realized on site to check the conformity of the lock with the hydraulic design criterions (installation of the gauges is illustrated in Figure 1).

Figure 1. Installation of the pressure gauges in the lock chamber.

Two opening laws of the valves were tested in order to decrease the cavitation and vibration problems (A: linear, B: bi-linear, faster at the beginning).

For different lock operations (filling, emptying, with one or two culverts, for the two opening laws tested), the following results are discussed: (1) the filling and emptying hydrographs; (2) the water slope in the lock chamber; (3) the wave amplitude in the downstream and upstream reaches; (4) the pressure, the vibrations and the sound level around the valves in order to quantify the cavitation level; and (5) the vibrations at different places of the structure.

The main conclusions are here summarized.

(1) The mean slope of the water surface in the lock chamber remains below the conservative value of 0.5‰. Nevertheless the lateral slope for an asymmetrical filling reaches 5‰. For this case, operational guidelines have to be proposed to avoid safety problems. The slopes measured for the opening law B are slightly higher than for the opening law A.
(2) When the locks Lanaye 3 and 4 are emptying in phase, the amplitude of the downstream wave is 10% higher than before the construction of the new lock. In other cases, the amplitude is lower.
(3) An intense level of cavitation is reached around the upstream and downstream butterfly valves. The opening law B allows decreasing the duration of cavitation ($\approx$3 min) against the law A. To have a better idea of the possible damages associated to cavitation, an inspection of the valve has to be planned after one year of use.
(4) Cavitation and vibration of the structures are two different problems. The measured vibrations are not dangerous for the structure. Nevertheless, a vibration having a specific frequency of 6 Hz was measured at the first floor of the control building in case of asymmetrical emptying. Pressure waves corresponding to this frequency have also been measured in the culverts. The origin of this pressure wave still has to be investigated. The vibration level and amplitude are similar for both opening laws.

Experimental investigation of the influence of breaking logs on the flow patterns induced by lock filling with gate openings

K. Verelst &, J. Vercruysse
Flanders Hydraulics Research, Antwerp, Belgium

P.X. Ramos & T. De Mulder
Hydraulics Laboratory, Civil Engineering Department, Ghent University, Ghent, Belgium

ABSTRACT

Navigation locks are key structures for the accessibility of ports and for navigation in canals and canalized rivers. Different systems exist for filling (and emptying) the lock chamber: short culverts bypassing the lock gates (with a stilling chamber), sidewall filling systems (with longitudinal culverts and ports), bottom filling systems and through the gate filling systems. For inland navigation locks in Belgium with a head limited to 2–3 m, through the gate filling systems are usually considered, i.e. lock filling takes place through openings in the lock gate sealed by some valves (often vertical lift valves). The latter levelling system is the subject of the research in this paper.

In the hydraulic design of a levelling system with openings in the lock gate, a compromise should be found between an acceptable lock filling time and acceptable forces on the moored vessels. One of the potential means to improve the energy dissipation of the filling jets and reduce the hydrodynamic forces on the ships, is to insert breaking logs (also referred to as energy dissipation bars) at the downstream side of the gate openings. These breaking logs are intended to enhance the spreading and energy dissipation of the filling jets.

To investigate the influence of breaking logs on flow patterns and energy dissipation in the lock chamber a dedicated (generic) physical model (presented in Figure 1) was built, consisting of one circular opening in a lock gate sealed by a vertical lift valve. Both configurations without and with breaking logs, having a square cross-section and placed vertically, were tested. The configurations with breaking logs differ with respect to the number and spacing of the breaking logs. The influence of the breaking logs on the discharge coefficient of the opening and on the flow velocity in the lock chamber was assessed.

Concerning the discharge coefficient of the opening and the flow velocity downstream of the lock gate, the physical model experiments show both logical (e.g. lower velocity in the lock chamber when

Figure 1. Physical model for studying the influence of breaking logs on flow pattern and energy dissipation in a lock chamber.

introducing breaking logs) and counter intuitive results (e.g. discharge coefficients for configurations with breaking logs being higher the than for configuration without breaking logs). The discharge coefficients derived from the physical model experiments are comparable to the ones derived from water level measurements in a Belgian inland navigation lock and to values reported in literature.

The velocity measurements and the visualisation of the flow pattern show at short distances from the gate opening the spreading of the jet and the velocity decay along the centreline. The length of the core of the jet is found to be significantly lower than the values reported in literature, although the comparison is hampered by differences in the definition of the (virtual) origin of the jet. Also an asymmetrical behaviour of the jet and backflow effects were noticed, due to the dimensions and downstream boundary conditions of the model.

The physical model experiments show some first observations of the influence of breaking logs on the flow pattern in the lock chamber and the jet behaviour. Nevertheless, a more intensive (i.e. on a finer grid) measurement of the flow velocity and a visualisation of the flow pattern in the lock gate itself will be carried out, in order to acquire a deeper understanding of the behaviour of the jet from an opening in the lock gate when breaking logs are present.

Relation between free surface profiles and pressure profiles with respective fluctuations in hydraulic jumps

J.D. Nóbrega & H.E. Schulz
Department of Hydraulics and Sanitary Engineering, School of Engineering at São Carlos, University of São Paulo, Brazil

M.G. Marques
Institute of Hydraulic Research, Federal University of Rio Grande do Sul, Brazil

ABSTRACT

Because the hydraulic jump (HJ) is a dissipative singularity, it is used for example in stilling basins downstream of spillways. Concerns about stilling basins are the possibility of cavitation and uplift of baffle blocks, which are mainly related to the pressure fluctuations, and the water depths, relevant for the design of the walls of stilling basins.

Because of the influence of the strong turbulence on both the deformation and breakup of the free surface and on the pressure field in HJ, it is expected that the free surface fluctuations may be correlated to the pressure fluctuations on the bottom of the flume.

Previous studies that have focused their attention on the investigation of the dynamics of the air-water interface were developed by Murzyn et al. 2007, Kucukali & Chanson 2008, Chachereau & Chanson 2011, Nóbrega et al. 2014. However, there is not a study that provides a comparison and quantitative analyses for the relationship between the free surface and the pressure characteristics in HJs.

Therefore, the aim of this paper is to compare statistical information of water depths, measured using ultrasonic sensor, and pressure, found in the literature. The ultrasonic sensor was positioned at different locations at the central line of the experimental flumes, for inflow Froude numbers (F_1) from 2.8 to 5.3.

The results were analyzed in terms of mean and extreme values (Fig. 1), standard deviation, fluctuation coefficient and skewness distribution.

The values of the standard deviations of both fluctuations reached their maxima in the region of the jump roller, decaying then to a reasonable constant value afterwards, when the flow turbulence is less intense. Moreover, towards the tail of the jump, the fluctuation coefficient and skewness stabilizes around a constant value.

This study also allows to suggest that the length of HJs may be quantified based on the decay of the statistical parameters, which is of difficult measurement by visual observations.

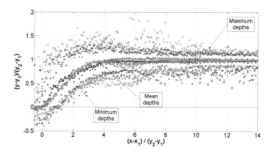

Figure 1. Distribution of mean and extreme water depth measurements (maximum and minimum) along hydraulic jumps with different F_1.

Therefore, the ultrasonic sensor has been proved to be a non-intrusive simple and promising technique for studies of strongly turbulent open flows. The study provides interesting information, showing similarities between the behavior of pressure and water depths evolution along HJs.

REFERENCES

Chachereau, Y. & Chanson, H. 2011. Free surface fluctuations and turbulence in hydraulic jumps. *Experimental Thermal and Fluid Science* 35(6): 896–909.

Murzyn, F., Mouzé, D. & Chaplin, J.R. 2007. Air-water interface dynamic and free surface features in hydraulic jumps. *Journal of Hydraulic Research* 45(5): 679–685.

Nóbrega, J.D., Schulz, H.E. & Zhu, D.Z. 2014. Free surface detection in hydraulic jumps through image analysis and ultrasonic sensor measurements. In *Hydraulic structures and society – Engineering challenges and extremes. 5th IAHR International Symposium on Hydraulic Structures*, Brisbane, 25–27 June 2014. Australia. doi:10.14264/uql.2014.42.

Kucukali, S. & Chanson, H. 2008. Turbulence measurements in the bubbly flow region of hydraulic jumps. *Experimental Thermal and Fluid Science* 33: 41–53.

Experimental study of head loss through an angled fish protection system

H. Boettcher, R. Gabl, S. Ritsch & M. Aufleger
Unit of Hydraulic Engineering, University of Innsbruck, Austria

ABSTRACT

An experimental study on head loss through an angled physical barrier for fish protection at hydropower plants was carried out. This is subject matter of a current research project on a new fish protection concept at hydropower intakes, the 'flexible fish fence'. A physical barrier is created by horizontally arranged steel cables. The structure is positioned upstream the turbine intakes, in a slight angle to the flow direction to guide fish to a bypass at the downstream end of the barrier (Brinkmeier et al. 2013).

Several studies of head loss at thrash racks or other hydropower intake structures have been published in the past (Kirschmer 1926; Meusburger 2002; Raynal et al. 2013; Kriewitz 2015). However, the special geometry of the flexible fish fence with slight angles and small bar spacings ($b \leq 30$ mm) as well as the flexible, vibrating structure of rough steel cables cannot be directly transferred to existing approaches for estimating head loss. A physical model on a scale of 1:2 was established to assess the impact of the geometry, defined by the angle and bar spacing, and velocity ranges on the head loss in the laboratory of hydraulic engineering, University of Innsbruck. The experiments of a preliminary study were performed for a physical barrier geometrically similar to the flexible fish fence, constituted of fixed horizontally arranged circular bar rods. Nine configurations with three angles to flow direction ($\alpha = 30, 45, 90°$, Fig. 1) and three bar spacings ($b = 5, 10, 15$ mm), resulting in blockage ratios of $p = 0.5, 0.33$ and 0.25, were studied for a range of different velocities. The screen was installed in a rectangular flume, which is 20 m long and 0.8 m wide. The Reynolds numbers in the channel varied from $Re = 750$ to 3000, whereby Re is related to the bar diameter $s = 5$ mm, Froude numbers are in the range of $Fr = 0.08$ to 0.3. The water depth was hold constant to 0.4 m.

As expected, the head loss Δh increased with Reynolds numbers and approach flow velocities. Furthermore, the results showed that the head loss coefficient ζ increased with increasing blockage ratio p and

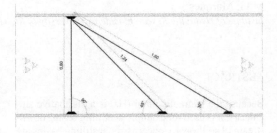

Figure 1. Layout of the studied angles of screen plane.

angle to flow direction α. Compared with estimated ξ-values according to Kirschmer (1926), Meusburger (2002) and Raynal et al. (2013), the best conformance with the measured data was observed with Raynal's approach. Thereby particularly the modified factor $k_\alpha = \sin^2(\alpha)$ proposed by Raynal et al. (2013) might better describe head loss through the studied angled structure for angles $\alpha \leq 45°$ than $k_\alpha = \sin(\alpha)$ proposed by Kirschmer (1926). Further experiments on head loss through the flexible fish fence are in progress. They are concentrating on the influence of vibrations and surface roughness on head loss, as a special feature of the steel cables.

REFERENCES

Brinkmeier, B., Boettcher & H., Aufleger, M. (2013). Flexible Fish Fences. *Proceedings of the 35th IAHR Congress.* Beijing.

Kirschmer, O. (1926). Untersuchungen über den Verlust an Rechen, *Mitteilungen Hydraulisches Institut München* Nr. 1.

Kriewitz, C.R. (2015). Leitrechen an Fischabstiegsanlagen: Hydraulik und fischbiologische Effizienz. *VAW-Mitteilungen* Nr. 230.

Meusburger, H. (2002). Energieverluste an Einlaufrechen von Flusskraftwerken, *VAW-Mitteilungen* Nr. 179.

Raynal, S., Chatellier, L., Courret, D., Larinier & M., David, L. (2013). An experimental study on fish-friendly trashracks – Part 1. Inclined thrashracks, *Journal of Hydraulic Research* 51(1): 56–66.

Scale effects for air-entraining vortices at pipe intake structures

K. Taştan
Department of Civil Engineering, Gazi University, Ankara, Turkey

ABSTRACT

Intake structures are used to withdraw water from reservoirs and rivers for several purposes such as irrigation, energy production, industrial cooling etc. If the vertical distance between intake and the water surface (submergence of the intake) is smaller than a certain limit value, which is called critical submergence, air starts to enter the intake due to free-surface air-core vortices. Since the entrained air reduces the efficiency of the intake and may generate operation problems downstream, determination of critical submergence and the influencing parameters is critically important. Hydraulic engineers have generally used Froude similarity between model and prototype intakes. However, previous studies showed that the Froude similarity may not represent the air entrainment or critical submergence at prototype intakes. Yıldırım and Kocabaş (1995) introduced the concept of imaginary critical spherical sink surface (CSSS) for the prediction of critical submergence of pipe intakes located in open channel flow, by superposing potential point sink flow and uniform flow. In this study, it is shown that if the ratios of average pipe intake velocity to the velocity at the CSSS between the model and prototype pipe intakes are identical (kinematic similarity), dimensionless critical submergences (ratio of critical submergence to the intake diameter) of corresponding intakes become also identical. Kinematic similarity concept can be applied both for the intakes located in still-water reservoir or cross-flow. In cross-flow cases, since the velocity at the CSSS is equal to the $U_\infty/2$ (if the intake is not affected by the friction caused by a boundary close to the free-surface or imposed circulation), pipe intakes having identical V/U_∞ also have identical S_c/D (in which, U_∞ is the average velocity of cross-flow, V is the average velocity at the intake, S_c is the critical submergence, and D is the diameter of the intake). Table 1 shows that kinematic similarity should be used instead of Froude similarity (F is intake Froude number).

In the case of still-water reservoir, the velocity at the CSSS cannot be predetermined. Therefore, tests should be conducted to obtain the velocity at the CSSS. This study showed that for similar flow conditions, velocities at the CSSS(s) related to model and prototype pipe intakes may be considered identical and kinematic similarity is still valid as shown in Table 2 (experimental results in Table 2 are taken from Jain 1977 and V_c is the velocity at the CSSS).

Table 1. Kinematic similarity and Froude similarity for vertical downward flowing pipe intakes in open channel flow (Taştan, 2016).

Similarity	D (cm)	F	V/U_∞	S_c/D
Kinematic similarity	3.6	7.52	25.64	1.83
	8.35	1.66	24.27	1.82
Froude similarity	3.6	2.18	7.52	1.03
	6.75	2.16	17.78	1.52

Table 2. Kinematic similarity and Froude similarity for vertical downward flowing pipe intakes in still-water (Taştan 2016 and Taştan and Yıldırım 2014).

Similarity	D (cm)	F	V/V_c	S_c/D
Kinematic similarity	2.70	8.75	647	9
	1.15	20	646	9
Froude similarity	7.46	2	249	5.57
	2.94	2	37	2.15

REFERENCES

Jain, A.K., 1977. Vortex formation at vertical pipe intakes. Ph.D. Thesis University of Roorkee at Roorkee, India.

Taştan, K. and Yıldırım, N., 2014. Effects of Froude, Reynolds and Weber numbers on an entraining vortex. Journal of Hydraulic Research, 52(3): 421–425.

Taştan, K., 2016. Scale and flow boundary effects for air-entraining vortices. Proceedings of the Institution of Civil Engineers- Water Management, DOI: 10.1680/jwama.15.00118.

Yıldırım, N. and Kocabaş, F., 1995. Critical submergence for intakes in open channel flow. Journal of Hydraulic Engineering, 121(12): 900–905.

Analysis of various Piano Key Weir geometries concerning discharge coefficient development

M. Oertel & F. Bremer
Hydraulic Engineering Section, Civil Engineering Department, Lübeck University of Applied Sciences, Lübeck, Germany

ABSTRACT

Piano Key Weirs (PKW) are relatively new hydraulic structures and research investigations start a couple of years ago. PKWs have modified the labyrinth weir geometry and guarantee an effective discharge, e.g. as flood release structure on a dam or as replacement of regular shape weirs in river systems. Current research focuses on discharge coefficients for various discharge amounts. For small discharges regular weirs are less effective than PKWs, hence replacements might be reasonable. Thereby, a couple of questions are in the main focus of interest, like e.g. general hydraulic phenomena, sediment transport, fish climb capability, or geometry adaption. Usually, PKWs are analyzed in scaled experimental models, while discharge coefficients can be calculated via Poleni or Du Buat approach. Additionally, numerical 3D CFD simulations are available to confirm experimental model results. The present paper deals with various laboratory-scaled PKW geometries and analyzes their upstream water surface profiles, discharge coefficients and efficiencies. Investigations are carried out within a 10 m long and 0.8 m wide tilting flume. Upstream water levels are measured via ultrasonic sensor as longitudinal water surface profiles.

It could be shown, that upstream water surface profiles are covered by three-dimensional flow effects due to the PKW's special geometry with inlet and outlet keys. For the inlet key water surface profiles will be accelerated close to the weir's longitudinal centerline and decelerated due to occurring resistance forces at structure's overflowing edges, which results in a waved profile. Discharge coefficients show a typical development with larger values for small discharges since the structure is more effective if a non-influenced fully three-dimensional flow occurs and the total centerline crest length becomes flow effective. For increasing discharges the flow changes with two-dimensional characteristics and discharge coefficients will be reduced. Analyzing and comparing discharge coefficients Cd and normalized discharge coefficients Cdw confirms non-practicability of Cd values for efficiency statements because these values are calculated with included centerline crest lengths.

Figure 1. Investigated PKW geometries, top: PKW_A, middle: PKW_B, bottom: PKW_A$_F$.

Since these lengths will change with changing PKW geometry a direct comparison of coefficients is only reasonable if using constant (channel) widths within the Poleni formula – resulting in so-called normalized discharge coefficients Cdw. With these best performances can be identified for PKW_A (symmetric overhang lengths) and PKW_AF (symmetric overhang lengths with included fractal elements) geometries. Further investigations are necessary to confirm negligence of scale effects; especially for PKW_AF geometry, where fractal elements are small sized.

Determination of discharge coefficient of triangular labyrinth side weirs with one and two cycles using the nonlinear PLS method

M.A. Nekooie
Department of Emergency Management, Malek Ashtar University of Technology, Tehran, Iran

A. Parvaneh
Department of Civil Engineering, Sharif University of Technology, Tehran, Iran

A. Kabiri-Samani
Department of Civil Engineering, Isfahan University of Technology, Isfahan, Iran

ABSTRACT

Side weirs are hydraulic control structures widely used in irrigation, drainage networks and waste water treatment plants. These structures can be served for adjustment and diverting of flow with minimum energy loss. In spite of many studies were carried out on rectangular side weirs, the studies on oblique and labyrinth side weirs are scarce.

Borghei et al. (2013) studied triangular labyrinth side weirs (Fig. 1). They conducted over 200 experimental tests (see Table 1) and introduced Eqs. (1) and (2) for estimation of the De Marchi coefficient of discharge C_M for triangular labyrinth side weirs with one and two cycles respectively.

$$C_M = \frac{0.076 \times \left(\frac{L'}{B}\right)^{1.935} + \left(\frac{p}{h_1-p}\right)^{0.240}}{1 + 0.802 \times \left(\frac{L'}{L}\right)^{-2.435} + 0.621 \times \left(\frac{Fr_1}{\sin(\delta/2)}\right)^{-0.403}} \quad (1)$$

$$C_M = \frac{-0.269 \times \left(\frac{L'}{B}\right)^{-1.188} + \left(\frac{p}{h_1-p}\right)^{0.180}}{1 + 0.649 \times \left(\frac{L'}{L}\right)^{-0.505} + 0.056 \times \left(\frac{Fr_1}{\sin(\delta/2)}\right)^{-1.275}} \quad (2)$$

Partial least square (PLS) is a robust method to estimate and fit multivariable statistical data. In effect, PLS is used to determine a dependent variable in terms of independent variables for nonlinear problems. Ramamurthy et al. (2006) used this method in hydraulic engineering for the first time. In this study, based on the experimental data from more than 210 laboratory tests and through using the multivariable nonlinear partial least square (PLS)

Table 1. The accuracy of Eqs. (20) and (21), compared with the presented equations by Borghei et al. (2013) for triangular labyrinth side weirs with one and two cycles respectively.

	Triangular Labyrinth Side Weir			
	One cycle		Two cycles	
Accuracy	Eq. (1)	Eq. (20)	Eq. (2)	Eq. (21)
Ave. Error	5.93%	5.81%	5.51%	5.64%
NRMSE	0.3423	0.3315	0.3638	0.3685

method, nonlinear Eqs. (20) and (21) are presented for discharge coefficient C_M of triangular labyrinth side weirs with one and two cycles. The obtained empirical equations relating C_M with the relevant geometric and hydraulic dimensionless parameters $L'/B, L'/L, Fr_1/\sin(\delta/2)$ and $p.\sin(\delta/2)/(h_1 - p)$ in a rectangular open channel. Comparison between results of the new presented equations and the measured data shows that the proposed empirical equations can predict the discharge of diverted flow over side weirs with good accuracy. Table 2 shows a better comparison between the accuracy of Eqs. (20) and (21) with Eqs. (1) and (2) in predicting the discharge coefficient C_M for triangular labyrinth side weirs.

REFERENCES

Borghei, S. M., Nekooie, M. A., Sadeghian, H., and Jalili, M. R. (2013). "Triangular labyrinth side weirs with one and two cycles." Proc. Inst. Civ. Eng., Water Manage, 166(1), 27–42.

Ramamurthy, A. S., Qu, J., and Vo, D. (2006). "Nonlinear pls method for side weir flows." J. Irrig. Drain. Eng., 132(5), 486–489.

Discharge capacity of conventional side weirs in supercritical conditions

A. Parvaneh
Department of Civil Engineering, Sharif University of Technology, Tehran, Iran

M.R. Jalili Ghazizadeh
Department of Water Engineering, Shahid Beheshti University, Tehran, Iran

A. Kabiri-Samani
Department of Civil Engineering, Isfahan University of Technology, Isfahan, Iran

M.A. Nekooie
Department of Emergency Management, Malek Ashtar University of Technology, Tehran, Iran

ABSTRACT

A side-weir is a hydraulic control structure widely used in irrigation and drainage networks and water treatment plants. As well as discharge adjustment, side weir can be served for preliminary treatment and decrease of the sediment load. The other advantages of side weir lies in its capability of diverting flood flows with minimum turbulence and energy loss. Only a few studies have been carried out on the supercritical flow over rectangular side weirs. The first equation was proposed by De-Marchi (1934) based on the assumption of constant specific energy along weir length. The supercritical flow over a side-weir is a typical case of spatially varied flow with decreasing discharge. Hager (1987) developed some equations to determine profile and discharge in the sub- and supercritical flows using modification of energy principle. He has presented Equation 1 for Froude number between 2 and 4.

$$C_M = 0.36 - 0.08 Fr_1 \quad (1)$$

He performed his tests on the side-weir with fixed length and limitation of $Fr_1 < 2$. In relation with supercritical flow, it can be mentioned to Hager (1994), and Crispino et al. (2015) for flow over circular channels Jalili Ghazizadeh (1994), Oliveto et al. (2001), and Ghodsian (2003) for flow over rectangular channels.

This study is aimed at investigating the variations of specific energy along the side-weir using experimental results. In the current tests, Froude number is between 1 and 2. Table 1 shows the variation range of the measured parameters.

Froude number varies within a range in which hydraulic jump can occur. Due to the importance of the current subject, the related tests to hydraulic jump and supercritical flow tests have been distinguished.

According to the obtained results, for the tests with upstream discharge of 45 lit/sec, the second-order relationship $Q/Q_1 = \alpha X^2 + \beta X + 1$ is valid

Table 1. Variation range of experimental data.

Run no.	L (cm)	p (cm)	S (%)	Q_1 (lit/s)	Fr_1
102	20–60	1–19	−0.5–2.5	35–100	1–2

between dimensionless parameters, namely Q/Q_1 and $X = x/B$. Therefore, $q = dQ/dx$ varies linearly with respect to X. Equations 2 and 3 have been developed for tests with the discharge of 45 lit/sec to determine α, β in terms of studied dimensionless parameters.

$$\alpha = -0.523 \frac{p}{y_1} - 0.770 \frac{L}{B} + 0.203 \quad (2)$$

$$\beta = 4.166 \frac{p}{y_1} + 0.042 \frac{L}{B} - 0.660 \quad (3)$$

REFERENCES

De-Marchi, G. (1934). "Sagio di teoria di fuzionamente degli stramazzi laterali" L'Elettrica, Millano, Italy, 11(11), 849–860.

Ghodsian, M. (2003). "Supercritical flow over a rectangular side weir", Canadian Journal of Civil Engineering, 30(3), 596–600.

Hager, W. H. (1987), "Lateral out flow over side weirs", J. Hydr. Engrg., ASCE, 113(4), pp. 491–503.

Hager, W. H. (1994), "Supercritical flow in circular-shaped side weir.", J. Irrig and Drain. Engrg., ASCE, 120, 1–12.

Jalili Ghazizadeh, M. (1994). "Experimental evaluation of side-weirs" M.Sc. thesis in Civil Engineering, Sharif University of technology.

Oliveto, G., and Biggiero, V., and Fiorentino, M. (2001). "Hydraulic features of supercritical flow along prismatic side weirs." J. Hydr. Research, 39(1), 73–82.

Discharge coefficient of oblique labyrinth side weir

A. Kabiri-Samani
Department of Civil Engineering, Isfahan University of Technology, Isfahan, Iran

A. Parvaneh
Department of Civil Engineering, Sharif University of Technology, Tehran, Iran

M.A. Nekooie
Department of Emergency Management, Malek Ashtar University of Technology, Tehran, Iran

ABSTRACT

Labyrinth side weir is one type of weirs, which can be used when the length of opening is limited. One of the advantages of labyrinth side weir is to increase the effective length of weir perpendicular to the flow and, therefore, diverting more discharge with the same flow depth and weir geometry (opening and height). Discharge coefficient should be determined to investigate the weir performance and estimate the discharge passing over the weir. In this paper, hydraulic performance of labyrinth side weir with asymmetric geometry has been experimentally studied (Fig. 1). The change to the geometry of ordinary labyrinth side weirs causes an increase in the effectiveness of the length of weir as being more in line with the streamlines. Thus, hydraulic behavior of this kind of labyrinth side weir with a constant opening length and different heights and angle has been investigated.

In this study, a labyrinth side weir with different geometry has been proposed. This change in the normal oblique side weir caused increase in both orthogonality length and effective length of side weir and consequently the discharge coefficient increased and side weir performance improved. Then through performing dimensional analysis, appropriate nonlinear relation for discharge coefficient were obtained

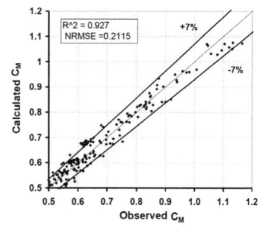

Figure 2. Observed C_M values versus measured C_M values from equation 1.

(Eq. 1). The discharge coefficients were within the range of $\pm 5\%$ (Fig. 2). Moreover, the results revealed that the deviated discharges by modified oblique side weir have increased about 17.8 % and 35.3 % for labyrinth and normal side weir, respectively.

$$C_M = \left[-0.18 \left(\frac{Fr_1}{\sin \theta'} \right)^{0.71} - 0.15(Fr_1)^{0.44} + \left(\frac{w}{y_1} \right)^{0.70} \right]$$

$$\times \left[-2.37 + 2.58 \left(\frac{w \sin \theta'}{y_1} \right)^{-0.36} \right] \quad (1)$$

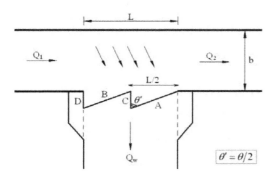

Figure 1. Plan view of asymmetric labyrinth side weir.

Reconstruction of a stage-discharge relation for a damaged weir on the Cavaillon river, Haïti

O. Carlier d'Odeigne, O. Roelandts, T. Verschoore, Y. Zech & S. Soares-Frazão
Institute of Mechanics, Materials and Civil Engineering, Université Catholique de Louvain, Louvain-la-Neuve, Belgium

ABSTRACT

Haiti is probably one of the most exposed countries to dramatic floods and inundations due, among other, to intense deforestation. Haiti is also extremely poor in hydrologic and hydrographic data, as this is not considered as a priority in a context of frequent disasters. The ambition of the program sustained by the Belgian Cooperation Administration (ARES-CCD) is to create, through an exemplary watershed and river reach along the Cavaillon River, a simple and repeatable methodology for enhancing hydrologic data and designing flood protection procedures.

One of the required data is the discharge of the river. As a weir, located near the village Dory, represents the upstream limit of the studied reach, we have the opportunity to evaluate the Cavaillon discharge from a continuous water level measurement located 50 m upriver. Unfortunately this infrastructure is in very bad condition as time and recurrent floods have done their work and considerably degraded the initial weir profile (Figure 1).

The aim of the presented work is to build a stage-discharge relation, considering that no data are available about the design and the construction of this weir and that the damages have the consequence that it is certainly not working in standard conditions. In situ survey, laboratory scale model and numerical modelling using the free Open Foam software are used to address the problem from complementary approaches.

To validate the numerical approach, a flow over a scaled model, 4 cm large and 7.5 cm height (crest level) has been studied. Laboratory measurements have been conducted by image analyses, allowing similar representations of the experimental observations and of the results obtained with the numerical model (Figure 2).

Figure 1. Illustrations of the Dory weir.

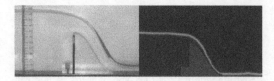

Figure 2. Scaled-model flow (right) and Open Foam simulation (left) related.

Figure 3. Dory weir model generated from in-situ survey.

In 2015 a dense geometrical survey has been conducted in Dory (Joseph et al., 2015), leading to the bathymetric data illustrated in Figure 3. The calibrated numerical model has been used to generate several flow scenarios on the spillway, considering its exact damaged geometry, in order to rebuild the local stage-discharge relation that could then be used as upstream boundary condition for the flow model.

Discussions about the results and the developed methodologies, based on field measurements, scaled model work and Open Foam software simulations will be detailed in the presentation.

REFERENCES

Bureau of Reclamation (1960). Design of small dams. A Water Resources Technical Publication. First Edition, 1960. Second Edition, 1973. Revised Reprint 1974.

Joseph A., Carlier d'Odeigne O., Zech Y., Gonomy N., Soares-Frazão S. (2015). Construction of one-dimensional model for the Cavaillon River, Haiti, from an in-situ survey of the bathymetry, e-proceedings of the 36th IAHR World Congress 28 June–3 July, 2015, The Hague, the Netherlands.

A two-fluid SPH model for landslides

A. Ghaïtanellis, D. Violeau, A. Leroy & A. Joly
EDF/LNHE & Saint-Venant Laboratory for Hydraulics, France

M. Ferrand
EDF/MFEE, France

ABSTRACT

A multi-fluid weakly compressible Smoothed Particle Hydrodynamics (SPH) method is used to model non-cohesive sediment transport. The main idea is to represent the sediment as a fluid with its own proper physical properties (density, viscosity, and so on). Then, a rheological law is used to model the non-Newtonian behaviour of the sediment.

The present SPH model is based on Hu and Adam's (2006) multi-fluid formulation and unified semi-analytical wall (USAW) boundary conditions (Ferrand et al., 2013). It is then able to handle accurately density and viscosity discontinuities at the interface of two immiscible fluids, guaranteeing the continuity of velocity and shear-stress across the interface. On the other hand, the USAW boundary conditions ensure an accurate pressure and shear-stress treatment at the wall even for complex boundary geometries. The non-Newtonian behaviour of the sediment is modelled using viscoplastic rheological model.

In this paper, the multi-fluid model is presented and tested on analytical and experimental cases. The chosen viscoplatic rheological law is then validated on a one-fluid theoretical case. The multi-fluid and rheological models are finally combined to simulate a schematic submarine landslide.

The experimental Assier-Rzadkiewicz's landslide case (Assier-Rzadkiewicz et al., 1997) is then studied. Results are compared with experimental data as

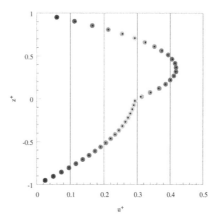

Figure 2. Two-fluid Poiseuille flow – Flow in a periodic pipe with density and viscosity ratio of $\rho_1/\rho_2 = \mu_1/\mu_2 = 0.25$ ($z^+ > 0 \Leftrightarrow$ fluid 1). The horizontal velocity profile at the steady state is plotted in color. The black dots represent the analytical solution.

well as numerical results (Capone et al., 2010) and a good agreement is obtained.

REFERENCES

Capone, T., A. Panizzo, & J. J. Monaghan (2010). Sph modelling of water waves generated by submarine landslides. *Journal of Hydraulic Research* 48(S1), 80–84.

Ferrand, M., D. Laurence, B. Rogers, D. Violeau, & C. Kassiotis (2013). Unified semi-analytical wall boundary conditions for inviscid, laminar or turbulent flows in the meshless sph method. *International Journal for Numerical Methods in Fluids* 71(4), 446–472.

Hu, X. & N. A. Adams (2006). A multi-phase sph method for macroscopic and mesoscopic flows. *Journal of Computational Physics* 213(2), 844–861.

Rzadkiewicz, S. A., C. Mariotti, & P. Heinrich (1997). Numerical simulation of submarine landslides and their hydraulic effects. *Journal of Waterway, Port, Coastal, and Ocean Engineering* 123(4), 149–157.

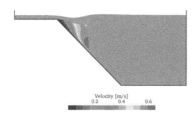

Figure 1. Submarine landslide – Snapshot of the simulation at $t = 0.4$ s.

Hydraulic modelling strategies for flood mapping. Application to coastal area in central Vietnam

Vo Ngoc Duong
Faculty of Water Resource Engineering, University of Science and Technology, The University of Da Nang, Viet Nam

Philippe Gourbesville
Innovative City lab URE 005, Polytech'Nice Sophia, Nice Sophia Antipolis University, France

ABSTRACT

Flood map nowadays is seen as an indispensable tool in urbanism, flood prevention and mitigation. For this reason, establishing flood map is mighty necessary for developing the socio economy of a river catchment. In recent years, creating this kind of map based on hydraulic models has been applied and proved good efficiencies in mitigating the consequences of flood catastrophes to human at many regions on the world. However, because of the lack of observed data in large catchments as well as developing countries, applying this work for these regions becomes a huge challenge for hydrologists. The insufficiency of meteo hydrological data, coarse resolution of topography, land cover data ... brings many difficulties to choose a suitable model or decide a reasonable model structure for flood modelling in these catchments. This study via the flood modelling process at downstream of Vu Gia Thu Bon catchment, a coastal region in Viet Nam central will compare the differences between 1D model, 2D model, Quasi 2D model and 1D/2D coupling model for flood simulation. The study also presents the uncertainties of input data such as topography, land use, rainfall, and boundary condition when modelling flood events. These simulations are carried out on the modules of Mike by DHI software: Mike 11, Mike 21, Mike Flood. The results might show strong and weak points of each model. These could help modelers to get several judgments when selecting the model to build the flood map. This study is expected to give some usefulnesses for flood modeling in the coastal part of a big catchment.

Estimation of 1D-confluence model parameters in right-angled discordant beds' confluences using 3D numerical model

D. Đorđević
Faculty of Civil Engineering, University of Belgrade, Belgrade, Serbia

I. Stojnić
(CERIS) – Instituto Superior Técnico – TU Lisbon, Portugal
LCH – EPFL, Lausanne, Switzerland

ABSTRACT

Extensive bathymetric surveys in river confluences during mid-1980s revealed that there was a difference in bed elevations between the tributary and main channels in the majority of surveyed confluences (Kennedy, 1984). The difference was created through a combined effect of the deposition of coarse sediment particles that had been arriving from upstream channel and deepening of the scour hole at the entrance to the post-confluence channel due to enhanced turbulence caused by collision of the combining flows (Constantinescu et al., 2011) The presence of the bed step at the tributary entrance to the confluence affects momentum transfer from the tributary to the main river. A proper estimation of this influence is of crucial importance for an accurate prediction (calculation) of upstream water levels (flow depths) in 1D flow modelling of dendritic river networks and lengthy river reaches. However, existing 1D-confluence models that were intended for the treatment of a confluence as an internal boundary condition in such analyses had been developed for the concordant beds' case, i.e. for the case when bed elevations of the combining channels are equal.

In models proposed after 1980 the contribution of the tributary flow to the momentum equation, which is written for the direction of the main-river flow, is taken into account either using the correction coefficient σ for the junction angle α (Hager, 1987, 1989 and Gurram et al., 1997), or by observing pressure difference between the opposite tributary walls near the confluence (Ramamurthy et al., 1988) or, by introducing mean cross-sectional value of the flow deflection angle in the downstream section of the tributary channel (Hsu et al., 1998). Since all models were derived for the concordant beds' confluences, it should be examined whether values of their parameters also hold for the discordant beds' confluences.

Parameters of 1D-confluence models for discordant beds' confluences are estimated in this paper using numerical simulation results obtained from a 3D finite-volume based model SSIIM2 that was successfully validated in concordant beds' confluences (Dordević, 2014). Confluences with the low, moderate and maximal observed bed elevation discordance ratio $\Delta z_T/h_d$ are analysed ($\Delta z_T/h_d = \{0.10, 0.25, 0.50\}$) for three characteristic hydrological scenarios defined by discharge ratio values $D_R = Q_{MR}/Q_d = \{0.250, 0.583, 0.750\}$. Here Δz_T stands for the difference in bed elevations between the tributary and main channels, h_d stands for the flow depth in the main river in the confluence, Q_{MR} is the discharge in the main-river upstream of the confluence and Q_d is the total downstream discharge. It is shown that: 1) the mean flow angle $\bar{\delta}$ approaches junction angle α with the increase in $\Delta z_T/h_d$, especially when $D_R = 0.250$, 2) the value of Hager's correction coefficient σ is not constant and 3) the contribution of the tributary flow to the 1D momentum equation is under predicted when either parameter $\bar{\delta}$ or σ is used for its estimation.

REFERENCES

Constantinescu, G., Miyawaki, S., Rhoads, B., Sukhodolov, A. and Kirkil, G. 2011. Structure of turbulent flow at a river confluence with momentum and velocity ratios close to 1: Insight provided by an eddy-resolving numerical simulation, *Water Resour. Res.* 47, 9W05507, doi:10.1029/2010WR010018.

Dordević, D. 2014. Can a 3D-numerical model be used as a substitute to a physical model in estimating parameters of 1D confluence models?, *Proc. 3rd IAHR Europe Congress*, Porto, 158–167.

Gurram, S.K., Karki, K.S., and Hager, W.H. 1997. Subcritical junction flow, *J. Hydraul. Eng.*, ASCE, 123, 5, 447–455.

Hager, W. H. 1987. Discussion of Separation zone at open-channel junctions, *J. Hydraul. Eng.*, ASCE, 113, 4, 539–543.

Hager, W. H. 1989. Transitional flow in channel junctions, *J. Hydraul. Eng.*, ASCE, 115, 2, 243–259.

Hsu, C.C., Lee, W.J. and Chang, C.H. 1998. Subcritical open-channel junction flow, *J. Hydraul. Eng.*, ASCE, 124, 8, 847–855.

Kennedy, B. 1984. On Playfair's law of accordant junctions. *Earth Surface Processes and Lanforms*, 9: 153–173.

Ramamurthy, A. S., Carballada, L.B. and Tran, D. M. 1988. Combining open channel flow at right angled junctions, *J. Hydraul. Eng.*, ASCE, 114, 12, 1449–1460.

Practical application of numerical modelling to overbank flows in a compound river channel

P.M. Moreta & A.J. Rotimi
School of Computing and Engineering, University of West London, Ealing, London, UK

A.S. Kawuwa
Faculty of Engineering and Physical Sciences, University of Surrey, Guildford, UK

S. López-Querol
Department of Civil, Environmental and Geomatic Engineering, University College of London, Ealing, London, UK

ABSTRACT

Accurate estimation of flow rate in channels is of enormous significance for flood prevention. Flooding occurs when the quantity of water flowing along a channel is higher than its carrying capacity. Hence the need for accurate prediction of river discharges during flood conditions to mitigate the impact, thereby saving lives and properties has drawn greater attention of researchers and engineers in recent times. There are numerous methods and approaches that can be employed to facilitate accurate estimation and prediction of discharge, conveyance and water surface level of rivers during overbank flow.

Compound sections formed by a river channel and floodplains, are used in river channels design to provide additional conveyance capacity during high discharge periods. When the overbank flow occurs, the flow in the river channel is affected by the momentum transfer between the main channel and floodplains, which modifies water levels and velocity distributions given by traditional methods. One-dimensional (1D) models using the Single Channel Method (SCM) and the Divided Channel method (DCM) have been proven to be not accurate enough in compound channel flows. New more advanced models have been shown to accurately estimate discharge flows and depth-averaged velocity distributions. Shiono and Knight (1991) developed a quasi-two dimensional model (based on lateral distribution method) to model conveyance in compound cross sections. This approach has been used in the Environment Agency's Conveyance Estimation System (CES). The two-dimensional SRH-2D (Sedimentation and River Hydraulics 2 Dimensional) model solves the depth-averaged Navier-Stokes equations in a discretized reach of a river. In this work, published field measurements of the River Main in UK (Martin and Myers, 1991; Myers and Lyness, 1994) are analysed via.

1D, CES and 2D modelling, in order to find a practical solution to give good predictions of water levels

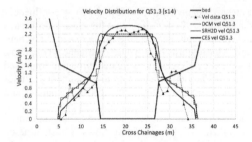

Figure 1. Velocities of field data (Martin and Myers, 1991) and computed with 1D (DCM), 2D (SRH) and CES.

and velocity distributions in overbank flows in river channels with floodplains.

The analysis and comparison of flow velocities has shown that the prediction of accurate velocity distributions in compound channel flow is a major challenge in numerical modelling. The 2D modelling trends to under predict main channel and floodplain interaction and slightly improve the depth-averaged velocities obtained with 1D model. In order to better simulate velocities, the CES is applied and gives a better representation of momentum interaction between main channel and floodplains (Figure 1). This methodology has been contrasted with field river data under gradually varied conditions, confirming previous and published results.

REFERENCES

Martin L.A. and Myers, W.R.C. (1991). "Measurement of overbank flow in a compound river channel". *Proc. ICE, Water Maritime and Energy*, 2, p. 645–657.

Myers, W.R. and Lyness, F.J. (1994). "Hydraulic study of a two-stage river channel". *Regulated Rivers Research and Management*, 9(4), p. 225–235.

Shiono, K., Knight, D.W. (1991). "Turbulent open channel flows with variable depth across the channel." *J. Fluid Mechanics*, 222, p. 617–646.

Comparison between different methods to compute the numerical fluctuations in path-conservative schemes for SWE-Exner model

Francesco Carraro, Valerio Caleffi & Alessandro Valiani
Dipartimento di Ingegneria, Universitá degli Studi di Ferrara, Ferrara, Italy

ABSTRACT

We consider the Shallow Water Equations (SWE) coupled with the Exner equation. To solve this non-conservative hyperbolic system, we implement a *P0P2-ADER* scheme [2], using a path conservative method for handling the non-conservative terms of the system. In this framework a comparison between three different Dumbser-Osher-Toro (DOT, see [3]) Riemann solver to approximate the Osher jumps at the cell edges $D^{\pm}$ is presented.

For a general formulation of the bedload transport flux, we define: *numerical* DOT the Riemann solver that computes $D^{\pm}$ using the numerical calculation of the eigensystem of the Jacobian matrix of the system; *analytical* DOT the one that adopts a new non-dimensional formulation of the analytical eigensystem proposed in [1]; *approximated* DOT the Riemann solver in which we introduced a new approximate estimation of the eigensystem for low-land rivers (i.e. with Froude number Fr $\ll$ 1) based on a perturbative analysis.

To test these different approaches a suitable set of test cases is used. Three of them are shown in this work: a test with a smooth analytical solution, a movable-bed Riemann Problem with analytical solution and a test in which the Froude number approaches unity. Finally a computational costs analysis is presented: Figure 1 shows the run durations with the three DOT Riemann solvers on varying the number of cells used to discretize the 1D space domain in a test case with smooth analytical solution.

From the comparison between the three different implementations, we obtained that the analytical DOT produce exactly the same results of the numerical DOT, but this latter method is much more resource demanding than the former. We notice that the efficiency increment is strictly related to the chosen quadrature rule to perform the integration along the path. Moreover, the computational cost of the numerical eigensystem should not be underestimated because it can grow very quickly with the number of interfaces between cells; specially for 2D or 3D models. Otherwise, the approximate DOT is the most efficient method of the three that we studied, but the limitations due to approximations must be considered for general problems. Quantitatively, using the approximate

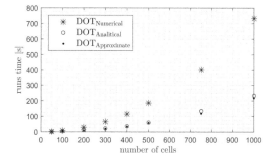

Figure 1. Computational costs analysis, i.e. durations of the runs with the three DOT Riemann solvers studied on varying the number of cells used to discretize the 1D space domain in a test case with smooth analytical solution.

DOT instead of the analytical one, the time saving results to be between 5% and 10%. However, because of the limitation due to the approximation, the advantage can be real only if: the spatial domain is very large, the simulation time is very long and the water flow is characterized from Fr $\ll$ 1. Thus, the tests shows that, in general, the use of analytical DOT for SWE-Exner model is recommended.

However, the approximate formulation of the eigensystem is much more easy than the analytical one, thus this new formulation can represent a starting point for further developments (e.g., the role of the gravity on the bottom topography and on the stability of dune's shape in sandy riverbeds).

REFERENCES

[1] Castro Díaz, M., Fernández-Nieto, E., and Ferreiro, A. Sediment transport models in Shallow Water equations and numerical approach by high order finite volume methods. *Computers & Fluids* 37, 3 (mar 2008), 299–316.

[2] Dumbser, M., Castro, M., Parés, C., and Toro, E. F. ADER schemes on unstructured meshes for nonconservative hyperbolic systems: Applications to geophysical flows. *Computers and Fluids* 38, 9 (2009), 1731–1748.

[3] Dumbser, M., and Toro, E. F. A Simple Extension of the Osher Riemann Solver to Non-conservative Hyperbolic Systems. *Journal of Scientific Computing* 48, 1–3 (2011), 70–88.

A new Osher Riemann solver for shallow water flow over fixed or mobile bed

D. Zugliani & G. Rosatti
Department of Civil, Environmental and Mechanical Engineering, University of Trento, Trento, Italy

ABSTRACT

The Osher solver [2] is a well-known numerical approach to estimate the solutions of the Riemann problems deriving from a finite volume method for hyperbolic systems of Partial Differential Equations (PDEs). Recently, in Dumbser & Toro [1], the applicability of the solver has been extended to purely non-conservative systems. Nevertheless, free surface flows are described by a system where both conservative and non-conservative terms are present. Some effort must be done to use the solver in this joined situation. In this work, we combine the conservative and non-conservative formulation ending up with a simple but powerful extension of the Osher solver suitable for the partially nonconservative PDEs systems. We also introduce a linear path in terms of primitive variables, instead of conserved ones. This approach reduces a little bit the computational cost in cases with simple linear relations between conserved and primitive variables (fixed-bed flows), while the cost reduction becomes more important when the relation is highly nonlinerar (mobile-bed flows) and the Jacobian of the fluxes can be expressed only in term of primitive variables. Finally, we exploited the possibility to use an explicit expression of the path integral of the non-conservative terms instead of a numerical approximation of it, e.g. the work of Rosatti & Begnudelli [3].

REFERENCES

Dumbser, M. and Toro, E.F., 2011. A simple extension of the Osher Riemann solver to non-conservative hyperbolic systems. Journal of Scientific Computing, 48(1–3): 70–88.

Osher, S. and Solomon, F., 1982. Upwind difference schemes for hyperbolic system of conservation laws. Mathematics of Computation, 38(158): 339–374.

Rosatti, G. and Begnudelli, L., 2010. The Riemann problem for the one-dimensional, free-surface shallow water equations with a bed step: Theoretical analysis and numerical simulations. Journal of Computational Physics, 229(3): 760–787.

Grid coarsening and uncertainty in 2D hydrodynamic modelling

V. Bellos & G. Tsakiris
Laboratory of Reclamation Works and Water Resources Management & Centre for the Assessment of Natural Hazards and Proactive Planning, School of Rural and Surveying Engineering, National Technical University of Athens, Greece

ABSTRACT

The aim of this paper is to study the impact of grid coarsening on the water depth results of a two-dimensional (2D) hydrodynamic simulation for both steady and dynamic evolution states. For the numerical simulation, the hydrodynamic model called FLOW-R2D which solves the full form of the 2D Shallow Water Equations through the Finite Difference Method and the McCormack numerical scheme (Tsakiris & Bellos 2014; Bellos & Tsakiris 2015).

Initially, the model is tested against the uncertainty introduced to the numerical results in a steady state scenario (hydraulic jump). Extending the uncertainty analysis to the dynamic state, an experimental test is employed (Soares-Frazao & Zech 2007) in which the flood wave created by a partial failure of a dam is propagated through a horizontal and dry bed with an isolated obstacle, which is oblique to the flow direction.

In both of these scenaria, three grid sizes were generated and compared, for the computational domain representation. For the uncertainty quantification, a grid convergence study is performed through the Grid Convergence Index (*GCI*) (Roache 1994). Besides, the observed order of convergence is also calculated. In the dynamic evolution scenario, apart from the *GCI* values, the relative error and the correlation between the numerical and the experimental results also estimated.

As far as the convergence study of the steady state is concerned, it is concluded that the uncertainty introduced to the numerical results derived by the FLOW-R2D model due to the grid coarsening, is relatively small. The simulations can be characterised as satisfactory, if the analytical solution and the numerical results derived for the three grid sizes are compared.

The observed order of convergence seems to have discrepancies from the formal order of convergence, probably due to the diffusion factor incorporated into the FLOW-R2D model which adds artificial viscosity.

Based on the convergence study of the dynamic evolution scenario, it seems that the uncertainty introduced to the numerical results in a complex computational domain with a dynamic evolution, is quite significant. Especially, this is observed during the first period up to the moment which the flood wave reaches to the gauge under inspection. As expected, the finest grid performs better than the others.

The observed order of convergence has significant discrepancies from the corresponding formal order of convergence. The negative values of the observed order of convergence indicate a non-monotonic behaviour, away from the asymptotic convergence. This fact has been also observed from other researchers in the past (Roy 2003).

It seems that despite of the fact that according to the correlation analysis the simulations can be characterised as satisfactory and that the water depths are simulated relatively accurately, the dynamic evolution of the phenomenon cannot be simulated quite successfully.

REFERENCES

Bellos, V. and Tsakiris, G., 2015. Comparing various methods of building representation for 2D flood modelling in built-up areas. Water Resources Management, 29(2), 379–397.

Roache, P.J., 1994. Perspective: a method for uniform reporting of grid refinement studies. Journal of Fluids Engineering, 116(3), 405–413.

Roy, C.J., 2003. Grid convergence error analysis for mixed-order numerical schemes. AIAA Journal, 41(4), 595–604.

Soares-Frazao, S. & Zech, Y. 2007. Experimental study of dam-break flow against an isolated obstacle. Journal of Hydraulic Research, 45, 27–36.

Tsakiris, G. and Bellos, V., 2014. A numerical model for two-dimensional flood routing in complex terrains. Water Resources Management, 28(5), 1277–1291.

Migration characteristics of a meandering river: The Madhumati river, Bangladesh

M.S. Banda
Department of Hydraulic Engineering, Leichtweiß-Institute (LWI), TU Braunschweig, Germany

S. Egashira
International Centre for Water Hazard and Risk Management (ICHARM), PWRI, Japan

ABSTRACT

Linearized 1-D models are widely used to simulate morphodynamic behavior of natural meandering rivers. Ikeda et al. (1981) developed a linear solution to predict near-bank excess velocity based on shallow water flow model. In their approach, bank erosion rate is linearly related to the near-bank excess velocity through an empirical erosion coefficient. Hasegawa (1989) proposed a bank migration model, which is based on the same linear relation as Ikeda et al.'s, but includes the erosion rate formula that is derived from integration of sediment continuity equation. This study aims to investigate whether these meander migration models are able to reproduce the observed behavior of a meander bend of the Madhumati River (Bangladesh).

Simulation results show that near-bank excess velocities are small in the upstream of bend apex, reach to the maximum value at bend apex and then decrease in the downstream. Correspondingly, the bank migration follows the same trend (Fig. 1). The agreement between predicted and observed migration is good in the upstream of bend apex, but the migration is underestimated in the downstream of bend apex. On the contrary, when an influence of the near-bank excess flow depth on the bank migration is added to Hasegawa's model, it is able to simulate the downstream migration well. Although the near-bank excess velocities and observed near-bank excess flow depths were out of phase, they together amplified the bank migration rate, which resulted in shifting the migration towards the downstream direction. However, this method necessitates careful estimation of flow width from asymmetrical cross-sections at bend. Nevertheless, such meander migration models provide a valuable and easy-to-use tool to analyze migration characteristics of meandering rivers.

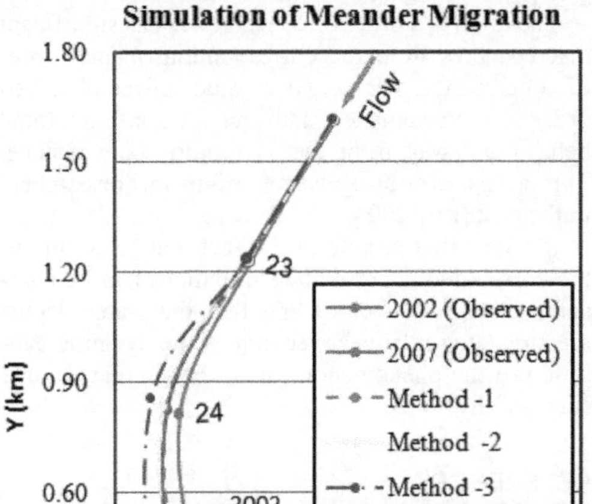

Figure 1. Simulation of meander migration – Madhumati River (Bangladesh): The digitized coordinates of the channel centerline in 2002 is used as the initial location for simulation. The distance between successive nodes is equal to one channel width. Method 1 is based on Ikeda et al. (1981). Method 2 & 3 follow Hasegawa's model. Channel centerlines as predicted in 2007 by method 1, 2 & 3 are shown with the sequence dashed (green), long-dashed (yellow) and long-dash-dotted (blue) lines. The predicted channel centerlines are compared with the observed location in 2007.

REFERENCES

Hasegawa, K., 1989. Universal bank erosion coefficient for meandering rivers, Journal of Hydraulic Engineering. ASCE 115(6), 744–765.

Ikeda, S., Parker, G. and Sawai, K., 1981. Bend theory of river meanders. Part 1. Linear development. J. Fluid Mech. 112, 363–377.

Employing surrogate modelling for the calibration of a 2D flood simulation model

V. Christelis, V. Bellos & G. Tsakiris
Laboratory of Reclamation Works and Water Resources Management & Centre for the Assessment of Natural Hazards and Proactive Planning, School of Rural and Surveying Engineering, National Technical University of Athens, Greece

ABSTRACT

In this paper, the problem of automatic calibration of the in-house FLOW-R2D hydrodynamic model is addressed. FLOW-R2D solves the full form of the 2D Shallow Water Equations (2D-SWE) through the Finite Difference Method and the McCormack numerical scheme (Tsakiris & Bellos 2014; Bellos & Tsakiris 2015). 2D flood modelling involves numerical simulations varying from several minutes to hours and the application of automatic calibration methods is limited due to the increased computational requirements. Thus, the dominant approach is the trial and error method which requires subjective decisions by the user of the model and constraints the efficient exploration of the parameter space.

To alleviate the excess computational burden from the automatic calibration of FLOW-R2D model, a surrogate-assisted optimisation algorithm is employed, namely the Multistart Local Metric Stochastic Radial Basis Function (MLMSRBF) (Regis & Shoemaker 2007). RBF surrogate models have been recently applied in water resources management and are considered very promising for deterministic simulations (Christelis & Mantoglou 2015). Two parameters of the FLOW-R2D model are calibrated, that is, the Manning's friction coefficient required for the friction model on the entire computational domain, and the effective slope of the upstream boundary cross-section, necessary for deriving the upstream boundary steady flow condition. The case study refers to the flood event created after the Tous dam break, which occurred in Spain in October 1982 (Alcrudo & Mulet 2007).

The optimal results from the MLMSRBF algorithm are compared to those obtained by a FLOW-R2D-based automatic calibration using an evolutionary algorithm. Both approaches are applied within a limited computational budget of 100 FLOW-R2D simulations. The objective function is the sum of the square error between the maximum water depth values recorded at 21 gauges located in the inundated area and the numerical results produced by the FLOW-R2D model.

The optimisation results indicate a significant improvement of the objective function score over the trial and error method applied in a previous work. Interestingly, the MLMSRBF algorithm produced similar results with the FLOW-R2D-based automatic calibration, both in terms of the objective function value and the identification of parameter values.

It should be noted though, that the Manning coefficient derived from the automatic calibration methods, takes significantly high values compared to those found in the related literature. It appears that model users should consider this black-box behaviour of automatic calibration methods in real world applications. Ongoing work includes the comparison of several conventional and surrogate-assisted optimisation methods, based on different computational budgets and larger dimensions, as well as a detailed assessment of the FLOW-R2D calibration for the aforementioned case study.

REFERENCES

Alcrudo, F. and Mulet J., 2007. Description of the Tous Dam break case study (Spain). Journal of Hydraulic Research, 45, 45–57.
Bellos, V. and Tsakiris, G., 2015. Comparing various methods of building representation for 2D flood modelling in built-up areas. Water Resources Management, 29(2), 379–397.
Christelis, V. and Mantoglou, A., 2015. Pumping optimization of coastal aquifers using radial basis function metamodels. Proc. 9th World Congress of EWRA, Istanbul, 10–13 June 2015. (e-proceedings).
Regis, R.G. and Shoemaker, C.A., 2007. A stochastic radial basis function method for the global optimization of expensive functions. INFORMS Journal on Computing, 19(4), 497–509.
Tsakiris, G. and Bellos, V., 2014. A numerical model for two-dimensional flood routing in complex terrains. Water Resources Management, 28(5), 1277–1291.

Estimating stem-scale mixing coefficients in low velocity flows

F. Sonnenwald
Department of Civil and Structural Engineering, University of Sheffield, Sheffield, UK

I. Guymer, A. Marchant & N. Wilson
School of Engineering, University of Warwick, Coventry, UK

M. Golzar & V. Stovin
Department of Civil and Structural Engineering, University of Sheffield, Sheffield, UK

ABSTRACT

Stormwater run-off typically contains and transports a wide range of pollutants, resulting in negative environmental effects. A current, joint, research programme is investigating the effects of heterogeneous vegetation distributions in stormwater ponds. To further develop this understanding, laboratory work has been carried out to quantify mixing (i.e. dispersion) around uniform emergent artificial vegetation in low velocity flows.

Ponds can have a variety of geometries and are often non-linear in shape, making them complex three-dimensional systems (Persson 2000). To simulate mixing processes within stormwater ponds, the complex geometries require that dispersion around vegetation be characterised at least longitudinally (D_x) and transversely (D_y).

Most laboratory systems are instead designed to collect cross-sectionally well-mixed longitudinal dispersion data, i.e. only D_x. Chikwendu (1986) presents an n-zone model relating D_y and D_x through the transverse variation of longitudinal velocity. This suggests that a characterisation of longitudinal dispersion and transverse variation of longitudinal velocity from a laboratory system designed to collect cross-sectionally well-mixed longitudinal dispersion data could be used to approximate transverse dispersion coefficients. This paper aims to do so.

A 15 m long by 30 cm wide recirculating narrow Armfield flume was fitted with uniform emergent artificial vegetation. Dye tracing was carried out and optimised velocity and longitudinal dispersion coefficients were calculated. Due to velocity probe limitations, Computational Fluid Dynamics (CFD) modelling was used to estimate the transverse variation of the longitudinal velocity. West et al. (2016) present a laboratory Laser Induced Fluorescence (LIF) wide flume experimental system to record solute trace data varying in time and transversely simultaneously. The downstream concentration profiles of West et al. (2016) were predicted using Armfield values of D_x and values of D_y made using Chikwendu (1986), combining the Armfield values of D_x with CFD generated velocity data.

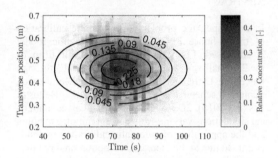

Figure 1. A downstream prediction (contours) compared to recorded downstream data. $R_t^2 = 0.7901$.

Figure 1 shows an example wide flume experimental downstream profile compared to a predicted downstream profile. Results show that the estimates of D_y made using Chikwendu (1986) with experimentally derived D_x and CFD generated velocity data are the correct order of magnitude and are fit-for-purpose for predicting mixing in a more realistic system.

ACKNOWLEDGEMENTS

This research was funded by EPSRC grants EP/K024442/1 and EP/K025589/1.

REFERENCES

Chikwendu, S. (1986). Calculation of longitudinal shear dispersivity using an n-zone model as n → ∞. *Journal of Fluid Mechanics 167*, 19–30.

Persson, J. (2000). The hydraulic performance of ponds of various layouts. *Urban Water 2*(3), 243–250.

West, P., J. Hart, I. Guymer, & V. Stovin (2016). Development of a laboratory system and 2D routing analysis to determine solute mixing within aquatic vegetation. In P. Rowinski and A. Marion (Eds.), *Hydrodynamic and mass transport at freshwater aquatic interfaces*. Springer.

SLIM: A model for the land-sea continuum and beyond

E. Deleersnijder
Université catholique de Louvain, Louvain-la-Neuve, Belgium
Delft University of Technology, Delft, The Netherlands

S. Blaise, P. Delandmeter, T. Fichefet, E. Hanert, J. Lambrechts, Y. Le Bars, V. Legat, J. Naithani, C. Pham Van, J.-F. Remacle, S. Soares-Frazão, C. Thomas, V. Vallaeys & D. Vincent
Université catholique de Louvain, Louvain-la-Neuve, Belgium

T. Hoitink & M. Sassi
Wageningen University, Wageningen, The Netherlands

V. Dehant & O. Karatekin
Royal Observatory of Belgium, Brussels, Belgium

E. Wolanski
James Cook University, Townsville, Australia

ABSTRACT

The Second-generation Louvain-la-Neuve Ice-ocean Model (SLIM, www.climate.be/slim_flyer) deals with the equations governing sea-ice, geophysical, environmental and groundwater phenomena by means of the (discontinuous Galerkin) finite element method on 1D, 2D or 3D unstructured meshes. To take advantage of state-of-the-art developments, SLIM is also being interfaced with existing tools (often based on radically different numerical methods), such as the well-known and widely used General Ocean Turbulence Model (www.gotm.net, GOTM). The post-processing of the results is achieved with the help of usual statistical and computer graphics methods. Other techniques are also resorted to, such as tracer and timescale methods derived from CART (Constituent-oriented Age and Residence time Theory, www.climate.be/cart) or network science tools (sites.uclouvain.be/networks) (Thomas et al. 2014). The hydrodynamics simulated by the aforementioned finite element model can be introduced into a number of SLIM-based environmental modules, which are capable of representing sediment transport (Delandmeter et al. 2015), as well as the fate of some classes of contaminants, namely microbiological pollutants (de Brauwere et al. 2014), endocrine disrupting compounds, heavy metals (Elskens et al. 2014) or radionuclides. A simple ecological model is being developed, whose aim is to simulate the evolution of various species of phyto- and zoo-plankton (Naithani et al. 2016).

SLIM has been applied successfully to a wide variety of standard, idealised test cases for geophysical and environmental fluid flows — including atmospheric ones (Blaise et al. 2015). It was seen that space-time mesh adaptivity pays off. Realistic problems were or are also dealt with, in particular, the application of SLIM to the Great Barrier Reef, Australia, and the land-sea continua of several rivers, namely the Scheldt (France, Belgium, The Netherlands), the Mahakam (Indonesia) and the Congo (Democratic Republic of the Congo). Finally, seas or lakes on some of the Jupiter and Saturn icy moons are being modelled (Vincent et al. 2016).

REFERENCES

Blaise, S., Lambrechts, J. and Deleersnijder, E., 2015. Stereographic projection for three-dimensional global discontinuous Galerkin atmospheric modeling, Journal of Advances in Modeling Earth Systems, 7: 1026–1050

de Brauwere, A., Gourgue, O., de Brye, B., Servais, P., Ouattara, N.K. and Deleersnijder, E., 2014. Integrated modelling of faecal contamination in a densely populated river-sea continuum (Scheldt River and Estuary), Science of the Total Environment, 468–469: 31–45

Delandmeter, P., Lewis, S.E., Lambrechts, J., Deleersnijder, E., Legat, V. and Wolanski, E., 2015. The transport and fate of riverine fine sediment exported to a semi-open system, Estuarine, Coastal and Shelf Science, 167: 336–346

Elskens, M., Gourgue, O., Baeyens, W., Chou, L., Deleersnijder, E., Leermakers, M. and de Brauwere, A., 2014. Modelling metal speciation in the Scheldt Estuary: combining a flexible-resolution transport model with empirical functions, Science of the Total Environment, 476–477: 346–358

Naithani, J., de Brye, B., Buyze, E., Vyverman, W., Legat, V. and Deleersnijder, E., 2016. An ecological model for the Scheldt estuary and tidal rivers ecosystem: spatial and temporal variability of plankton, Hydrobiologia (in press)

Thomas, C.J., Lambrechts, J., Wolanski, E., Traag, V.A., Blondel, V.D., Deleersnijder, E. and Hanert, E., 2014. Numerical modelling and graph theory tools to study ecological connectivity in the Great Barrier Reef, Ecological Modelling, 272: 160–174

Vincent, D., Karatekin, O., Vallaeys, V., Hayes, A.G., Mastrogiuseppe, M., Notarnicola, C., Dehant, V. and Deleersnijder, E., 2016. Numerical study of tides in Ontario Lacus, a hydrocarbon lake on the surface of the Saturnian moon Titan, Ocean Dynamics (in press)

*Hydrometeorological extremes, uncertainties
and global change*

Hybrid downscaling and conditioning for characterizing multivariate flooding extremes

M. del Jesus, P. Camus & I.J. Losada
Environmental Hydraulics Institute "IH Cantabria" Universidad de Cantabria, Santander, Spain

ABSTRACT

In standard hydrologic applications, the return period of any variable (river discharge, for instance) is directly related to the return period of its driver (in this case precipitation). This assumption greatly simplifies the characterization of derived magnitudes, for which there is normally no measured record and whose behavior must be numerically modeled. However, this simplification may lead to significant errors when the dynamics analyzed are the composition of different factors, i.e. when the magnitude is multivariate.

In this work we present a method that makes use of data from Atmosphere-ocean general circulation models (AOGCMs) to analyze large-scale climate variability (long-term historical periods, future climate projections) and to obtain a series of representative synoptic states on which to condition measured values. The proposed method can be considered a hybrid approach, which combines a probabilistic weather type downscaling model with a stochastic weather generator component. Predictand distributions are reproduced modeling the relationship with AOGCM predictors based on a physical division in weather types (Camus et al. 2014).

The multivariate dependence structure of the predictand (extreme events) is introduced linking the independent marginal distributions of the variables by a probabilistic copula regression (Ben Alaya et al. 2014). This hybrid approach is applied for the conditioning of daily precipitation and maximum sea water level to AOGCM data in the isle of Tenerife, Spain. Results are then used to obtain the set of hydraulic boundary conditions needed to determine the water level reached under flood conditions in a storm stream outlet.

Making use of weather types as conditioning variable for distribution fitting allows to capture very complex behaviors by means of the combination of simple and well-known parametric distributions. Simple parametric forms are used to fit conditioned

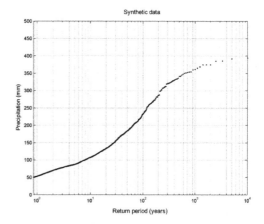

Figure 1. Empirical extreme regime of precipitation computed from synthetic data.

variables while complex shapes are obtained by means of the combination of the different conditioned fits.

Hybrid downscaling is also a useful tool to optimally balance the need for long time series for extreme regime characterization and the total computational cost of the operation. The maximum dissimilarity algorithm is also an important part of the procedure as it allows an adaptive approach to achieve the desired error tolerances.

REFERENCES

Ben Alaya, M. A., F. Chebana, & T. B. M. J. Ouarda (2014). Probabilistic Gaussian Copula Regression Model for Multisite and Multivariable Downscaling. *Journal of Climate* 27(9), 3331–3347.

Camus, P., M. Menéndez, F. J. Mendez, C. Izaguirre, A. Espejo, V. Cánovas, J. Pérez, A. Rueda, I. J. Losada, & R. Medina (2014). A weather-type statistical downscaling framework for ocean wave climate. *Journal of Geophysical Research: Oceans* 119(11), 7389–7405.

Application of artificial neural networks in meteorological drought forecasting using Standard Precipitation Index (SPI)

S. Golian, P. Yavari & H. Ruigar
Civil Engineering Department, Shahrood University of Technology, Iran

ABSTRACT

Drought is one of the natural and climatic disasters and causes abundant damages to human life and natural ecosystems worldwide every year. This study aims to compare the results of meteorological drought forecasting using non-linear auto regressive (NARX) neural network model with those of commonly-used multi-layer Perceptron (MLP) network for Shahrood climate station, located in North-East part of Iran. Different combination of input-output sets with various variables as inputs and SPI with different lag-time as output were tested to determine which combination has the best performance in prediction of future droughts. SPI drought index with 3 and 6 months aggregation period were applied in this study.

In general, both MLP and NARX networks had satisfactory performance in prediction of SPI index, but NARX has slightly better performance compared to MLP network for all the models. Also predicting SPI6 resulted in better performance compared to SPI3. Results showed that for both models, using temperature, precipitation, humidity, SPI3, SPI6 and SPI12 as input variables and SPI6 with one month lag as output has better performance compared to other input-output combination sets. The optimum calculated performance criteria were CNS (Nash-Sutcliff) = 0.954 and 0.962, R (correlation coefficient) = 0.989 and 0.986, MAE (mean of absolute error) = 0.033 and 0.036, RMSE (root mean square error) = 0.041 and 0.043 for training and test periods, respectively for NARX network.

However the use of climate variables or drought indices alone as the input parameters of the ANN models showed less accurate results compared to the input combination with both meteorological and SPI variables (in both MLP and NARX, models 3 and 4 had better results compared to models 1 and 2). As shown in the paper, NARX network has slightly better results compared to MLP network in all models. For the MLP network, when prediction lead-time increased, for all model and SPI with different aggregation time scales, the accuracy of predictions was reduced. But there was no such regular results among the outputs of the NARX network.

Overall, in prediction of SPI Index, SPI6 had better results compared to SPI3 with slightly better correlation coefficient. This could be as the result of this fact that SPI6 aggregates the previous 6-month precipitation and has smoother fluctuations compared to SPI3 which aggregates the 3-month precipitation; hence, this make the prediction of SPI6 more precise in comparison with SPI3 Also, both neural networks used in this study, i.e. MLP and NARX, were enough accurate to predict meteorological drought using appropriate inputs. Hence, they are recommended for the same arid areas. With regard to importance of input variables, it is suggested that the effects of some parameters such as sea level temperature, evapotranspiration etc. are also investigated in drought forecasting to improve the model performance. Furthermore, the other drought indices like MSDI, HDI, Palmer etc. can be used in drought forecasting.

Accuracy assessment of ISI-MIP and FAO hydrological modelling results in the Upper Indus Basin

Asif Khan
Department of Engineering, University of Cambridge, UK
Department of Engineering, University of Engineering and Technology, Peshawar, Pakistan
Internationational Institute for Applied System Analysis, Laxenburg, Austria

Mujahid Khan
Department of Engineering, University of Engineering and Technology, Peshawar, Pakistan

ABSTRACT

Millions of people rely on river water originating from snow and ice melt in the Hindukush-Karakoram-Himalayan (HKH). One such basin is the Upper Indus Basin (UIB), where snow and glacier melt contribution is more than 80%, therefore is highly susceptible to global warming and climate change. Accuracy of available hydro-climatic studies' results are vital for future precise policy making and sustainable water resource development. Therefore, this research evaluates accuracy of various ISI-MIP and FAO hydro-climatic studies results, during 1985–1998 and 1961–1990 respectively, for six sub-basins of the UIB. This research evaluates accuracy of bias corrected five GCMs' precipitation data sets, input of ISI-MIP hydrological models, and CRU data, input of FAO hydrological model, all based on basin-wide mass balance assessment. First the input precipitation data sets have been evaluated followed by comparison of modelled and measured flows. Basin-wide mass balance assessment show that all precipitation data sets used in ISI-MIP and FAO hydrological models significantly underestimate precipitation in the UIB, particularly in the Karakoram sub-basins. All ISI-MIP (6) and FAO hydrological models provide consistent but significantly low modelled flows (<50%) as compared to the measured records in all sub-basins, except for the Kharmong basin. FAO water scarcity shows severe water scarce conditions in the UIB. FAO and ISI-MIP under-estimated modelled flows (and water scarcity) are artefacts of use of under-estimated precipitation data use. This study shows that results of ISI-MIP and FAO are not true representative of hydro-climatic conditions in the UIB, therefore cannot be used in precise and accurate policy making and water resource management.

A Gaussian design-storm for Mediterranean convective events

I. Andrés-Doménech, R. García-Bartual, M. Rico Cortés & E. Albentosa Hernández
Universitat Politècnica de València (UPV), Instituto de Ingeniería del Agua y Medio Ambiente (IIAMA), Valencia, Spain

ABSTRACT

The Spanish Mediterranean façade usually experiences flash floods caused by torrential rainstorms. Consequently, achieving a deeper knowledge about rainfall characteristics has been of paramount importance to improve design criteria and to develop more efficient and safer hydraulic infrastructures. This research develops a new framework to define design storms with temporal and spatial distribution, based on observed rainfall patterns. It is applied to the rainfall features of Valencia, Spain. Convective episodes have been identified from the 1990–2012 period and individual convective storms have been extracted (Llasat, 2001); statistical description of their internal characteristics (volume, duration and peak intensity) has been performed. Dependencies between the three variables are explored and mathematical relationships proposed. The processed data is finally used to fit the theoretical model for the design storm, based on a Gamma function reproducing temporal patterns (García-Bartual and Marco, 1990), and a non-dimensional Gaussian function to model the spatial distribution of convective rain cells (Rodríguez-Iturbe and Eagleson, 1987):

$$i(r, t) = i_0 g(r) f(t) \quad (1)$$

where r is the distance to the storm centre, t is time and i_0 is the maximum storm intensity. $f(t)$ is the non-dimensional function defining the temporal pattern of the storm and $g(r)$ is the non-dimensional function defining its spatial distribution.

Once the model formulated and its main properties known, parameters of $f(t)$ are fitted according to the 73 storms available in the data set and theoretical properties of the model validated.

The proposed design storm fits accurately the observed temporal pattern of historic convective events (Figure 1). Parameters of $f(t)$ have been successfully characterised from the data set and their relationships with the other main descriptors of the storm (volume and duration) established and validated.

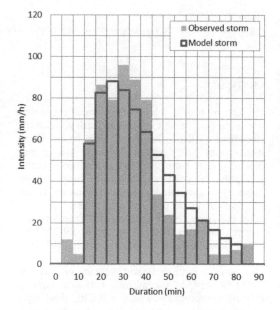

Figure 1. Model fit for the storm on 11th October 2007.

Ongoing work is now being developed to completely define the practical steps for the proposed methodology: $g(r)$ parameters estimation from radar data and a convenient procedure for the storm return period estimation.

REFERENCES

García-Bartual, R. and Marco, J. 1990. A stochastic model of the internal structure of convective precipitation in time at a raingauge site. *Journal of Hydrology*. 118: 129–142.

Llasat M.C. 2001. An objective classification of rainfall events on the basis of their convective features: application to rainfall intensity in the northeast of Spain. *International Journal of Climatology*. 21: 1385–1400.

Restrepo-Posada, P. J. and Eagleson, P. S. 1982. Identification of independent rainstorms, *J. Hydrol.* 55 (1–4): 303–319.

Rodríguez-Iturbe, I. and Eagleson, P. S. 1987: Mathematical

Extreme hydrological situations on Danube River – Case study Bezdan hydrological station (Serbia)

M. Urošev
Geographical Institute "Jovan Cvijić" SASA, Belgrade, Serbia

I. Leščešen
Department of Geography, Tourism and Hotel Management, Faculty of Science, University of Novi Sad, Novi Sad, Serbia

D. Štrbac
Geographical Institute "Jovan Cvijić" SASA, Belgrade, Serbia

D. Dolinaj
Department of Geography, Tourism and Hotel Management, Faculty of Science, University of Novi Sad, Novi Sad, Serbia

ABSTRACT

Hydrological extremes (floods and hydrological droughts) are natural hazards that are not limited to specific regions, but occur worldwide and therefore, impact a very large number of people. Extreme hydrological situations on Danube River at Bezdan for the 1931–2014 period were analyzed. Usually these two hydrological extremes are analyzed separately, our goal was to define them in similar way, which allows us to compare the results of frequency analysis. Floods and hydrological droughts were defined as events over and below selected threshold levels respectively. The selected thresholds were Q_5 and Q_{10} for floods, and Q_{90} and Q_{95} for droughts. Two variables, volume and duration of floods and droughts were analyzed using frequency analyses.

The volumes and durations of floods and droughts were analyzed by annual maximum (AMS) and partial duration series (PDS) methods. Parameter estimation for distributions was calculated using method of L-moments. In flood frequency analysis of the AMS Log-Pearson 3 distribution has the best agreement with empirical data for both volumes and durations, while for hydrological droughts it was the Pearson 3 distribution for Q_{90} and Log-Pearson 3 for Q_{95} threshold. In flood frequency analyses of the PDS joint distribution of Poisson and Weibull fits the empirical distribution of largest volumes and durations, while for hydrological droughts best distribution depends on selected thresholds. Comparing the results of all 16 generated series we can conclude that, in general, both AMS and PDS models give satisfactory estimates of return periods of flood (Fig. 1) and drought volumes for return periods smaller than 100 years, where for larger return periods AMS model performs better. AMS provides better estimates of return periods of flood durations than PDS model for both Q_5 and Q_{10} thresholds, while

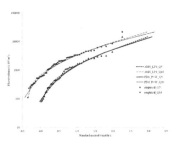

Figure 1. Comparison between AMS and PDS model for flood volume frequency analysis for thresholds Q_5 and Q_{10}.

it's opposite for drought durations – PDS give better results for both Q_{90} and Q_{95}.

The frequency analysis pointed out the most extreme hydrological events on Danube River at Bezdan for 1931–2014 period – the 1965 flood, being the largest flood both in terms of volumes and durations, and 1947 and 1954 hydrological droughts.

Definition of floods and drought through its volumes and durations is more useful to hydrological engineers than a single value of instantaneous water discharge as it was highlighted for 2013 flood. Two practical applications of frequency analyses of flood and drought volumes and durations for Danube at Bezdan, were pointed out at the end of paper.

REFERENCES

Madsen, H., Rasmussen, P., & Rosbjerg, D. 1997. Comparison of annual maximum series and partial duration series methods for modeling extreme hydrological events. 1. At-site modeling. *Water Resources Research*, 33(4): 747–757.

Tallaksen, L.M., Madsen, H. & Clausen, B. 1997. On the definition and modeling of streamflow drought duration and deficit volume. *Hydrological Science Journal*, 42(1): 15–33.

Presenting an empirical model for determining the sugar beet evapotranspiration by GDD parameter (Case study: Torbat-Jam, Iran)

F. Jan Nesar, A. Khashei Suiki, S.R. Hashemi & S. Moradi Kashkooli
Department of Water Engineering, Faculty of Agriculture, University of Birjand, Iran

ABSTRACT

So far, various methods were presented for determining the reference crop evapotranspiration in different parts of the world. The most popular and prestigious of them can be combined methods of Penman family, modified Blany-Criddle, Hargreaves – Samani and thornthwaite. Due to the lack of the lysimeter data in many parts of Iran, presenting an equation according to the regional condition with high precision is so paramount. The aim of this paper is presenting a method with higher precision for determining the evapotranspiration of the sugar beet in Torbat-Jam by using the meteorology and *GDD* (Growing Degree Day) parameter and determined methods were evaluated with FAO- Penman- Monteith method as a standard model for determining the evapotranspiration. RMSE, R^2 indexes, and Nash-Sutcliffe index, NS, were used for comparing fitness indexes.

$$ET_C = \frac{p}{n} + \frac{n}{T} + (-21.82) + \frac{n}{N} + GDD \quad (1)$$

$$ET_C = GDD \times 0.175 + RH \times (-0.00609) + \frac{n}{N} \quad (4)$$

$$ET_C = (\frac{p}{n}) \times 80.47 + n^{1.0861} + T \times (-0.000206)$$
$$+ \frac{n}{N} \times (-13.4085) + GDD \times 0.023 + RH \times (-0.00377) \quad (8)$$

Table 1. The amounts of correlation coefficient, error and NS of each method.

equation	R^2	RMSE	NS
1	0.693	14.082	0.99
2	0.651	1.169	0.515
3	0.719	58.97	−1229
4	0.678	1.117	0.558
5	0.64	1.183	0.503
6	0.64	1.048	0.611
7	0.47	4.147	−5.08
8	0.64	1.21	0.99
9	0.683	1.05	0.99
(10) Blany-Criddle	0.657	1.22	0.99

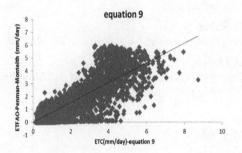

Figure 1. The correlation lines between every equation and FAO- Penman- Monteith equation.

Results showed that number four equation is appropriate for calculating the water requirement of crop. So R^2, RMSE indexes and Nash-Sutcliffe coefficient were 0.683, 1.117 and 0.99, respectively for equation 9.

REFERENCES

Kaviani M, Kaviani A, Taheri M (2013) Application of Sebal Algorithm in estimating the actual amount of evapotranspiration in Ghazvin plains and comparing them with lysimeter data.

Mohsen Movahed SA, Akbari M, Dadivar M, Khodshenas M A (2013) Comparing and evaluating some models of evapotranspiration estimating of reference crop in Arak.

Rahimi khoob A (2006) Presenting a model to convert the evapotranspiration of pan to the evapotranspiration of reference crop. Case study: Khuzestan province. National conference on irrigation and drainage management network. The University of Ahvaz.

Impact of rainfall variability on the sewerage system of Casablanca city, Morocco

Leila Ennajem & Dalila Loudyi
Water and Environmental Engineering, University of Hassan II Casablanca,
Faculty of Science and Technology of Mohammedia, Morocco

ABSTRACT

Climate change is a global fact that affects the future of many communities. Many floods and droughts were recorded during the last two decades. In Morocco, as in the whole Mediterranean region, many cities have been affected by floods generated by extreme precipitation event (Mohammedi, 2002, Algiers, 2006, Casablanca and Rabat, 2010 and 2012, E'ttaref, 2012) that caused considerable damage to the urban and industrial infrastructure. Moreover, climate change suggests a change in the cycle of rainfall in Morocco along with an increase in the average annual temperature, precipitation, intensity and frequency of heavy rainfall until mid-century.

In this article, the particular case of the city of Casablanca, which is considered among the most vulnerable coastal cities to natural hazards and extreme weather phenomena such as extreme precipitation, will be presented. Projections for 2030 describe a decrease of 6% to 20% of annual rainfall and a warming of 1.3°C for the city of Casablanca. However, it is also predicted that warmer temperatures and low rainfall will be accompanied by more frequent and intense extreme precipitation events, providing additional flow that will increase the pressure already faced by urban drainage systems and causing substantial damages to the urban infrastructure and significant economic losses. Many structural and non-structural measures have been taken by the city sewerage managers to adapt the network to the new changes in precipitation and discharge evacuated by sewers to the Atlantic coast. The impact of these measures on the performance of the network and therefore their economic impact on the city will also be presented.

REFERENCES

Driouech, F., R. Sebbari, A. Mokssit, A. Nmiri, M.I. Omar, A. El-Hadidi, D. Bari, (2009): Scénarios de changement climatique au niveau des villes côtières d'Afrique du nord: Alexandrie, Casablanca, Rabat et Tunis. Projet de la Banque Mondiale «Adaptation au changement climatique et aux désastres naturels des villes côtières d'Afrique du Nord».

Egis BCEOM International/IAU-IDF/BRGM, 2011a. Phase 1: Évaluation des risques en situation actuelle et à l'horizon 2030 pour la ville de Casablanca. Projet: Adaptation au changement climatique et aux désastres naturels des villes côtières d'Afrique du Nord. World Bank report, GED 80823T, 261 p.

Implications of CMIP5 derived climate scenarios for discharge extremes of the Rhine

F.C. Sperna Weiland, M. Hegnauer & H. Van den Boogaard
Deltares, Delft, The Netherlands

H. Buiteveld & R. Lammersen
Rijkswaterstaat, Lelystad, The Netherlands

J. Beersma
KNMI, De Bilt, The Netherlands

ABSTRACT

In 2013 the IPCC published the 5th assessment report based on a new generation of global climate model simulations forced with representative concentration pathways (RCP's). From these data KNMI constructed a set of four climate scenarios – the KNMI'14 scenarios (van den Hurk et al. 2014). In this study, the scenarios tailored to the Rhine basin are used to assess the potential effects on discharge extremes in the river Rhine and their potential implications for flood protection design in the Netherlands.

In the Netherlands the current method for flood protection design requires the derivation of discharge extremes for return periods of 1000 up to 30,000 years. Since the available observed discharge records are too short to derive such extreme return discharges, the extreme value analysis is based on very long synthetic discharge time-series generated with the Generator of Rainfall And Discharge Extremes (GRADE).

GRADE consists of a stochastic weather generator for the current climate and hydrological- and hydrodynamic models to simulate discharge. To derive the probability distributions of extreme discharges under climate change the synthetic rainfall and temperature series simulated with the weather generator are transformed following the KNMI'14 climate scenarios. The hydrological model is forced with the synthetic meteorological time-series after which the propagation of the flood waves is simulated with the hydrodynamic model.

The analysis shows that all KNMI'14 scenarios project increases in discharge extremes for the river Rhine under future climate conditions hereby posing increased requirements for flood protection design to maintain the current safety standard.

REFERENCES

Buishand, T.A. & Brandsma, T. 2001. Multi-site simulation of daily precipitation and temperature in the Rhine basin by nearest-neighbor resampling. Water Resources Research 37: 2761–2776.

Hurk, B. van den, Siegmund, P., Klein Tank, A. (Eds), Attema, J., Bakker, A., Beersma, J., Bessembinder, J., Boers, R., Brandsma, T., van den Brink, H., Drijfhout, S., Eskes, H., Haarsma, R., Hazeleger, W., Jilderda, R., Katsman, C., Lenderink, G., Loriaux, J., Van Meijgaard, E., Van Noije, T., Van Oldenborgh, G.-J., Selten, F., Siebesma, P., Sterl, A., de Vries, H., Van Weele, M., de Winter, R. & van Zadelhoff, G.-J. 2014. *KNMI'14: Climate Change scenarios for the 21st Century – A Netherlands perspective*. KNMI-publication WR-2014-01, 26/5/2014, pp. 120, KNMI, De Bilt, The Netherlands.

Van Pelt, S.C.., Beersma, J.J, Buishand, T.A., Van den Hurk, B.J.J.M. & Kabat., P. 2012. Future changes in extreme precipitation in the Rhine basin based on global and regional climate model simulations. Hydrology and Earth System Sciences 4: 4517–4530, doi:10.5194/hess-16-4517-2012.

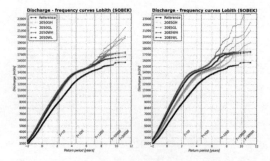

Figure 1. Discharge – frequency curves for the Rhine at Lobith for all scenarios and years, based on the hydrodynamic model results (Sobek), with the effect of upstream flooding. In grey the results without the correction for the flood areas along the German stretch upstream of Lobith are shown.

How will be future rainfall IDF curves in the context of climate change?

H. Tabari, P. Hosseinzadehtalaei & P. Willems
Department of Civil Engineering, Hydraulics Division, KU Leuven, Belgium

S. Saeed, E. Brisson & N. Van Lipzig
Department of Earth and Environmental Sciences, KU Leuven, Leuven, Belgium

ABSTRACT

The design statistics for water infrastructures are typically derived from rainfall intensity–duration–frequency (IDF) curves which compound frequency and intensity aspects of rainfall events for different durations. Current IDF curves are constructed based on historical time series, with an underlying temporal stationarity assumption for the probability distribution of extreme values. However, climate change casts doubt on the validity of this assumption due to ongoing and projected changes in the intensity and frequency of extreme rainfall. In this study, IDF curves for historical periods obtained from the convection permitting CCLM model with spatial and temporal resolutions of 2.8 km and 15 minutes and an ensemble of climate models (CMIP5) are validated based on observations-based curves. After this validation, future climate IDF relationships are obtained based on a quantile perturbation approach.

It is concluded that the sub-hourly precipitation intensities at 15 and 30 minutes in the IDF curves derived from the CCLM 2.8 km model underestimate the observed extreme rainfall intensities. For the daily intensities, less deviation is observed for both the CCLM and the CMIP5 GCM runs.

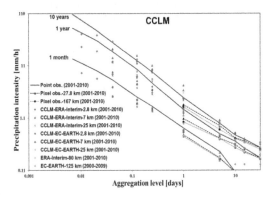

Figure 1. Baseline IDF curves using the CCLM model and its driving GCM and reanalysis versus corresponding IDF curves obtained using point and pixel observations at Uccle for the historical period 2001–2010.

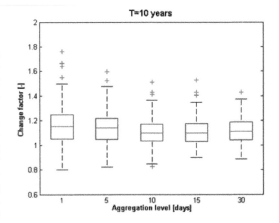

Figure 2. Relative changes in IDF relationships between the control (1961–1990) and scenario (2071–2100) runs of the CMIP5 GCMs for a 10-year return period and aggregation levels, combining all RCPs.

Future climate projections show potentially strong changes in extreme rainfall intensities, making the historical climate based IDF design standards unsuitable for the future extreme events.

The results of this study indicate that the current IDF curves are not sufficient to represent future precipitation patterns and emphasize the necessity of upgrading the curves for designing, operating and maintaining municipal water management infrastructures in the future.

REFERENCES

Mailhot, A., Duchesne, S., Caya, D., Talbot, G. 2007. Assessment of future change in intensity–duration–frequency (IDF) curves for southern Quebec using the Canadian Regional Climate Model (CRCM). Journal of Hydrology 347: 197–210.

Tabari, H., Taye, M.T., & Willems, P. 2015. Water availability change in central Belgium for the late 21th century. Global and Planetary Change 131: 115–123.

Willems, P., & M. Vrac, 2011: Statistical precipitation downscaling for small-scale hydrological impact investigations of climate change. Journal of Hydrology 402: 193–205.

Influence of model uncertainty on real-time flood control performance

E. Vermuyten, P. Meert, V. Wolfs & P. Willems
Department of Civil Engineering, KU Leuven, Leuven, Belgium

ABSTRACT

Worldwide, the risk of flooding increases due to climate change and urbanization trends, among other reasons. In densely populated areas like the Flanders region of Belgium, limited space is available to install additional flood control infrastructure like retention basins. Therefore, it is important to strive for improved management based on existing infrastructure.

Previous research on the Demer basin in Belgium (Van den Zegel et al. 2014, Vermuyten et al. 2014, Chiang & Willems 2015, Vermuyten et al. 2015) has shown that existing infrastructure can be used in the most optimal way by means of real-time control. An advanced control algorithm has been developed for that purpose by combining Model Predictive Control (MPC) with a Reduced Genetic Algorithm (RGA). In order to limit the required optimization time, a conceptual river model was developed. For this, the semi-automated calibration approach designed by Wolfs (2015) was used. Damage-water level functions were added to the model to compare the efficiency of different control strategies based on economic considerations (reduced flood damage cost). The developed approach significantly reduced the total economic flood damage cost.

In this previous research, the meteorological inputs (rainfall forecasts) and the river models were considered perfect. No uncertainties in the inputs and models were taken into account. The current research is the first step in quantifying these uncertainties, studying how they can be taken into account during the MPC-RGA optimization process and analyzing their influence on real-time flood control efficiency.

This research focuses on one of these uncertainties, namely the model uncertainty. First, this uncertainty is quantified and the influence of the length of the prediction horizon and the initial river model conditions are investigated. Furthermore, the loss in real-time control performance due to this uncertainty and a method to reduce this loss is examined. This method consists of a data assimilation technique whereby only the current observations are used to update the river model states. Preliminary results show that even a basic data assimilation method is capable of reducing the loss in performance for small model uncertainties.

REFERENCES

Chiang, P., Willems, P. 2015. Combine evolutionary optimization with Model Predictive Control in real-time flood control of a river system. *Water Resources Management* 29(8): 2527–2542.

Van den Zegel, B., Vermuyten, E., Wolfs, V., Meert, P. and Willems, P. 2014. Real-time control of floods along the Demer river, Belgium, by means of MPC in combination with GA and a fast conceptual river model. *11th International Conference on Hydroinformatics*, New York City, USA, 17–21 August 2014.

Vermuyten, E., Van den Zegel, B., Wolfs, V., Meert, P. and Willems, P. 2014. Real-time flood control by means of an improved MPC-GA algorithm and a fast conceptual river model for the Demer basin in Belgium. *6th International Conference on Flood Management*, Sao Paulo, Brazil, 16–18 September 2014.

Vermuyten, E., Meert, P., Wolfs, V. and Willems P. 2015. Using a fast conceptual river model for floodplain inundation forecasting and real-time flood control – a case study in Flanders, Belgium. *21st International Congress on Modelling and Simulation*, Broadbeach, Queensland, Australia, 29 November – 4 December 2015.

Wolfs, V., Meert, P., Willems, P. 2015. Modular conceptual modelling approach and software for river hydraulic simulations. *Environmental Modelling & Software* 71: 60–77.

Flash flood prediction, case study: Oman

E. Holzbecher, A. Al-Qurashi, F. Maude & M. Paredes-Morales
German University of Technology in Oman, Sultana of Oman

ABSTRACT

In Oman flash floods regularly cause damages on buildings, infrastructure and cost lives. Prediction of flood routing can thus be a valuable tool to mitigate the effects of extreme precipitation events. These tools can be part of warning systems and used for risk maps in urban planning. However, the task is challenging due to severe uncertainties, lack of hydrological parameters as well as frequent occurrence and importance of hydrological extremes. While flood routing is general a difficult task in arid and semi-arid climates, the problem is aggravated in Oman, not only due to geographical conditions, but also due to urbanization etc (Al-Qurashi 2013).

In 2015 the project entitled 'Towards a flood-resilient Omani society: improved tools for flood management' has started working to investigate the problem of flash floods in Oman. The general objective of the research is to develop improved methods for flood risk management. This includes on one hand improving the scientific knowledge for analysing and assessing flood hazards and flood risks, and on the other hand developing a set of decision support tools for flood risk management as well as a public flood information platform. The project is funded by The Research Council (TRC) of Oman.

In the project three watersheds were selected for detailed investigation. These are the wateersheds Ma'awil, Bani Kharus and Al Fara, located in the South Batinah region. Data from several observation points with rainfall and run-off measurements, operated regularly by the Ministry of Regional Municipalities and Water Resources (MRMWR), are available. Other relevant data are obtained from other Omani agencies and ministries.

We describe the approach for flood prediction using 2D distributed physically-based integrated modelling. For flood routing we solve the 2D shallow water equations. After some benchmarking we chose the ANUGA code (Roberts et al. 2015) for integrated hydrological and hydraulic modelling. Our model of the stream channels of Al Fara watershed accounts for flow from different tributaries, friction and transmission losses. With this approach we are able to simulate the main features of run-off, observed when cyclone Phet hit the Omani coast in June 2010.

REFERENCES

Al-Qurashi, A. 2013. An overview on flood studies in Oman, in: UNESCO, Int. Seminar on Natural Disaster Risk Reduction, Proc.: 25–35.
Roberts, S., Nielsen, O., Gray, D., Sexton, J., Davies G. 2015. ANUGA User Manual, Release 2.0, Geoscience Australia.

Equipping the TRENT2D model with a WebGIS infrastructure: A smart tool for hazard management in mountain regions

N. Zorzi, G. Rosatti & D. Zugliani
Department of Civil, Environmental and Mechanical Engineering, University of Trento, Trento, Italy

A. Rizzi & S. Piffer
Trilogis Srl, Rovereto, Italy

ABSTRACT

Mountain regions are naturally exposed to extreme floods and geomorphic flows. In the last years, climate change has worsened this hazard exposure. Therefore, incisive actions and strategies are always more undeniable to safeguard urbanized areas.

In recent years, several mathematical models simulating hyperconcentrated flows and debris flows have been developed, with increasing reliability. Some of them have already been used to plan hazard protection and mitigation strategies, with encouraging results (see for instance Rosatti et al. 2015). However, the increasing trustworthiness of the most advanced models implies higher complexity and larger computational burdens, with a greater request of high-performing hardware. These drawbacks hinder a widespread diffusion of "best practices and best available technologies not entailing excessive costs" in risk-management field, despite the requirements contained in the UE Flood Directive (2007/60/EC). Therefore, new modelling solutions should be found to support the diffusion of cutting-edge models.

In this work, a smart and easy-to-use solution is proposed and applied to the TRENT2D model (Armanini et al. 2009, Rosatti and Begnudelli 2013), which is a state-of-the-art 2D model, simulating debris flows and hyperconcentrated flows over a mobile bed with a two-phase approach, reproducing erosion and deposition processes properly. For the purpose of enhancing and simplifying the use of the model, TRENT2D was converted into a service, according to the SaaS (Software as a Service) approach, and equipped with WebGIS technology. The new infrastructure, called TRENT2D WG, is hosted by a cloud server and offers a complete, flexible and user-friendly working environment, which allows to apply the model through the World Wide Web. This solution introduces several advantages in comparison with standalone software logic. It allows model input and output data to be managed, displayed and analysed straightforwardly in the same working environment where the model is applied, with an intuitive web Graphic User Interface (GUI). Model computational burdens are transferred from the user hardware to a high-performing server, so the model can be used almost independently of the properties of the accessing device. Moreover, maintenance is centralised and model accessibility is enhanced, bringing innovation closer to models users.

Furthermore, the infrastructure was enriched also with a GIS-based BUWAL-type (Heinimann et al. 1998) hazard-mapping procedure, accessible from the same GUI. In this way, model results can be used easily to assess debris-flow hazard levels, avoiding work fragmentation and encouraging the diffusion of advanced models and effective hazard-management practices.

REFERENCES

Armanini, A., L. Fraccarollo, & G. Rosatti (2009). Two-dimensional simulation of debris flows in erodible channels. *Computers & Geosciences* 35(5), 993–1006.

Heinimann, H. R., K. Hollenstein, H. Kienholz, B. Krummenacher, & P. Mani (1998). Methoden zur Analyse und Bewertung von Naturgefahren. Technical Report 85, Bundesamt für Umwelt, Bern.

Rosatti, G. & L. Begnudelli (2013). Two-dimensional simulation of debris flows over mobile bed: enhancing the TRENT2D model by using a well-balanced Generalized Roe-type solver. *Computers & Fluids* 71, 179–195.

Rosatti, G., N. Zorzi, L. Begnudelli, & A. Armanini (2015). Evaluation of the Trent2D model capabilities to reproduce and forecast debris-flow deposition patterns through a back analysis of a real event. In G. Lollino (Ed.), *Engineering Geology for Society and Territory – Volume 2, Acts of: International Association of Engineering Geology and the Environment IAEG XII Congress*, pp. 1629–1633. Springer International Publishing.

Reservoir operation applying a discrete hedging rule with ensemble streamflow prediction to cope with droughts

S. Lee & Y. Jin
Civil Engineering, Pukyong National University, Busan, Republic of Korea

ABSTRACT

Reservoir operations applying a hedging rule may be a useful method to cope with droughts. Hedging operations of a reservoir was studied by the three steps: a discrete hedging rule derivation, ensemble streamflow prediction, and reservoir operations applying the derived hedging rule curves. The hedging stages consist of concern, caution, alert, and severe that correspond to required flow release, instreamflow reduction, agricultural water reduction, and municipal water diminishing, respectively. A mixed integer programming was used to decide trigger volumes of available water for rationing that is the current storage plus the inflow volume of the future time period. The decided hedging rule curves of Hapcheon Dam are shown below. The Box-Whisker plots are depicted from the historical data.

The future inflow volume was computed by ensemble streamflow prediction using historical precipitation scenarios and precipitation forecast. The tank model with soil moisture structure was used to simulate ensemble streamflow. Reservoir operations of Hapcheon Dam, Republic of Korea were simulated applying the derived hedging rule curves and ensemble streamflow predictions. The results showed the reduction of maximum and overall water supply deficit.

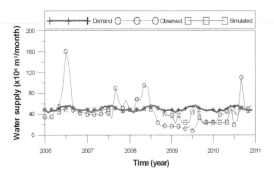

Figure 2. Water supply resulted from the discrete hedging rule and ensemble streamflow prediction.

Table 1. Critical values of monthly water supply.

Item	Observed	Simulated
Minimum water supply (10^6 m^3/month)	8.17	20.79
Maximum deficit for the minimum water supply (10^6 m^3/month)	46.20	34.16
Occurrence	July 2009	June 2009

The derived hedging rule curves are certainly worth practical reservoir operation for water supply, because they can be easily understood by dam operators under the proper future inflow forecasting such as the ensemble streamflow prediction.

REFERENCES

Shih, J., and Revelle, C., 1995. Water supply operations during drought: a discrete hedging rule. European Journal of Operational Research, 82, 163–175.

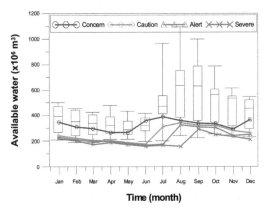

Figure 1. Derived hedging rule curves of Hapcheon Dam.

Determination of the interaction between surface flow and drainage discharge

S. Kemper, A. Schlenkhoff & J. Balmes
University of Wuppertal, Wuppertal, Germany

ABSTRACT

Managing urban flooding caused by heavy rainfall events requires new design approaches concerning the drainage system as well as the temporary surface water runoff. Latest developments on bidirectional coupled models, 1D-1D as well as 1D-2D models, are still employed in practice. After Butler & Davies (2011) these models are capable of representing the interaction of the major and minor systems under extreme flow conditions. Connecting elements between the surface and the underground drainage system are the so called street inlets or gullies – offered in different construction types and designs. Depending on the longitudinal and transversal slope of the street as well as the street inlet type the hydraulic efficiency of grate inlets (typically used in Germany) is hardly available, thus, physical model test runs were done to receive the requested information (Kemper & Schlenkhoff, 2015). Due to steep longitudinal slopes up to 10%, only supercritical flow conditions occur with flow depths up to 3 cm and flow velocities of approximately 1–2 m/s. The intercepted flow varies between $Q_I = 2.95$ l/s and $Q_I = 16.00$ l/s within the investigated model test runs (Figure 1).

Figure 1. Physical Model.

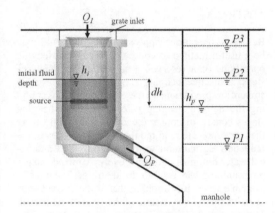

Figure 2. Underground drainage system, described in DIN 4052-2.

With three-dimensional numerical model test runs, the capacity of the underground system (Figure 2) to cope with the intercepted surface flow is analyzed. Several conditions – clogging effects as well as different pressure heads in the pipe system – were investigated. Based on the results from the numerical model test runs, a chart was developed to identify the maximal possible inflow from the surface to the underground system. With inflow flow rates up to 20 l/s (regardless the pressure head in the pipe system) the water can be drained in total with only less backwater effects in the inlet but no water entering the surface. Even with clogged sludge buckets of approximately 60 % flow rates up to 20 l/s (pressure head P1) can be drained without causing surface flooding.

REFERENCES

Butler, D., Davies, J.W. 2011. Urban Drainage. Third Edition. Spon Press.
Kemper, S., Schlenkhoff, A. 2015. Determination of the hydraulic efficiency of intake structures like grate inlets and screens in supercritical flow. E-proceedings of the 36th IAHR World Congress, The Hague, the Netherlands.

Effects of flow orientation on the onset of motion of flooded vehicles

C. Arrighi & F. Castelli
Department of Civil and Environmental Engineering, University of Florence, Italy

H. Oumeraci
Leichtweiss Institute for Hydraulic Engineering and Water Resources, Technical University of Braunschweig, Germany

ABSTRACT

Flood risk in urban areas is recognized as a crucial issue as the population living in cities is constantly increasing with a consequent rise in the value of exposed asset. It has been demonstrated that, in developed countries, most of the fatalities during a flood occurs inside vehicles. In fact, the combination of water depth and velocity acts so that a car can be swept away also for very low water depth. This work aims at identifying the critical conditions for vehicles incipient motion in order to support flood risk mapping and management and to promote people's education. The dimensionless mobility parameter θ_V derived in a previous study is modified to account for the effect of different flow orientations. The geometry and flood parameters used to define θ_V are water depth H, velocity U (i.e. Froude number), angle of flow incidence β and length L, height H_V and elevation h_c of the chassis of the vehicle, as represented in Figure 1.

3D numerical simulations in the OpenFOAM framework have been carried out to understand the role of the angle of incidence of the flow with respect to the car. The hydrodynamic forces are evaluated for different flow regimes, both subcritical and supercritical. The results of the simulations highlight that not only the area exposed to the flow matters for the onset of motion, but also the flow field under the vehicle planform. Existing flume experiments on small-scale models are used for the numerical model validation and as a reference for the interpretation of the numerical results.

REFERENCES

Arrighi, C., Alcèrreca-Huerta, J.C., Oumeraci, H., & Castelli, F. 2015. Drag and lift contribution to the incipient motion conditions of partly submerged flooded vehicle. Journal of Fluids and Structures 57: 170–184.

Shu, C., Xia, J., Falconer, R. & Lin, B., 2011. Estimation of incipient velocity for partially submerged vehicles in floodwaters. Journal of Hydraulic Research, 49: 709–717.

Xia, J., Falconer, R.A., Xiao, X. & Wang, Y. 2014. Criterion of vehicle stability in floodwaters based on theoretical and experimental studies. Natural Hazards, 70: 1619–1630.

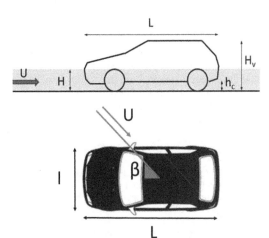

Figure 1. Definition sketch of the flood and vehicles characteristics accounted for in the mobility parameter θ_V.

Development of a computationally efficient urban flood modelling approach

V. Wolfs, V. Ntegeka, D. Murla Tuyls & P. Willems
Department of Civil Engineering, Hydraulics Division, KU Leuven, Leuven, Belgium

ABSTRACT

Urban flooding causes worldwide signification disruption to society, huge economic losses and serious health risks. The rapidly increasing urbanization and climate change will increase the risk of and damage due to urban floods. Increasing the resilience of urban areas to local rainfall-induced floods is therefore a major objective of present and future water management. Besides the design of adaptation measures to reduce the flood hazard, real-time urban flood forecasting and warning systems become increasingly important. Both the development of future-proof adaptation strategies and warning systems require models that can accurately simulate floods and the response of the urban drainage system with very short calculation times.

Due to the inherent complexity of the urban floods and the sewer system, mathematical models are required for decision making. Spatially very detailed 1D-2D hydrodynamic models exist that simulate urban floods, but these models suffer from (overly) complex and rigid model structures. Arguably even more importantly, simulating floods requires very long calculation times, impeding their use for numerous applications. Therefore, a new conceptual modelling approach was developed for simulating urban floods that is compatible with the framework developed by the authors for conceptual and efficient hydraulic modelling of rivers (Wolfs et al., 2015) and sewers (Wolfs and Willems, submitted). The methodology is highly computationally efficient and well-suited for applications requiring very fast simulation models, such as early warning systems and optimization, e.g. to determine real-time storage operations. The approach simulates the water levels and timing of urban floods using a modular structure that consists of a serial connection of artificial neural networks. After simulating the flood level, the flood extent can be visualized by means of GIS calculations using high-resolution digital terrain models. The approach is highly flexible and allows the modeller to focus on flood-prone areas of interest or

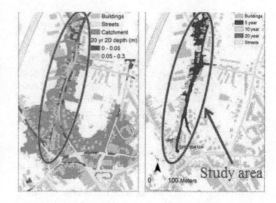

Figure 1. Maximum flood extent simulated by the InfoWorks-ICM for the 20-year storm (Left) and by the GIS-based approach for 5-, 10-, 20-year storms (Right).

on the dominating processes. The method can deal with different temporal and spatial scales, and can for instance use (weather) radar data as input.

A case study is performed on the sewer system of the city of Ghent. Urban flooding is simulated in a detailed 1D-2D InfoWorks ICM model. A conceptual model was then created using the simulation results of the detailed model as calibration data, and the results were compared to those of the detailed model. Both visual comparison and goodness-of-fit criteria show that the conceptual model can accurately emulate the results of the detailed 1D-2D ICM model, while the calculation time is an order of magnitude 10^6 times shorter than the ICM model.

REFERENCES

Wolfs, V., Meert, P., Willems, P., 2015. Modular conceptual modelling approach and developer for River quantity simulations. *Environmental Modelling and Software* 71: 60–77.

Wolfs, V., Willems, P., submitted. Modular conceptual modelling approach and software for sewer hydraulic computations. *Advances in Water Resources*.

Rapid flood inundation modelling in a coastal urban area using a surrogate model of the 2D shallow water equations

L. Cea, M. Bermudez, J. Puertas, I. Fraga & S. Coquerez
Environmental and Water Engineering Group, Departamento de Métodos Matemáticos y de Representación, Universidade da Coruña, Spain

ABSTRACT

Two dimensional shallow water models have demonstrated good capabilities for flood inundation mapping in urban areas. However, even if High Performance Computing techniques have greatly decreased the computational time needed to run a 2D inundation model, this approach remains unsuitable for applications as real time forecasting or uncertainty propagation in a Monte Carlo context, which require fast urban inundation models. One possibility to meet this requirement are data driven models, which directly relate input and output data without analyzing the physical processes involved. Artificial intelligence techniques have been applied in the field of rainfall-runoff modelling at the catchment scale (Dawson et al., 2006), but very few applications to flood inundation modelling have been investigated (Pender & Liu, 2011).

In this paper we propose and compare the application of different linear, non-linear and non-parametric regression techniques as surrogate models of the 2D shallow water equations (SWE) applied to flood inundation mapping in ungauged urban areas. A coastal urban area is used as a test case. Regression models were developed to predict the maximum water depth computed by a SWE model at selected control points (Fig. 1). The predictor variables were derived from the discharge and tide data prescribed at the open boundaries of the SWE model domain, being several variable combinations tested.

Regression based on least-squares support vector machines resulted in better water depth estimates than standard linear regression. The best performance was achieved when considering the discharge of the three input tributaries and the tide level data as predictor variables. In this way, mean absolute errors represented less than 5.5% of the water depth variation computed by the SWE model at all control points. Overall, the results show the potential of the proposed regression technique for fast and accurate computation of flood inundation maps.

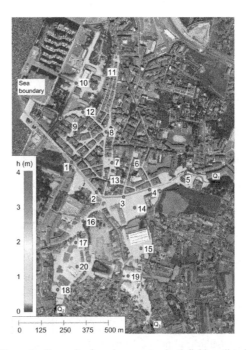

Figure 1. Example of maximum water depth field predicted by the SWE model in the urban area. Location of control points (1–20) and open boundaries (Q_1, Q_2, Q_3 and sea boundary).

REFERENCES

Dawson, C.W., Abrahart, R.J., Shamseldin, A.Y. and Wilby, R.L, 2006. Flood estimation at ungauged sites using artificial neural networks. Journal of Hydrology, 319, pp.391–409.

Pender, G., Liu, Y. 2011. Rapid Flood Inundation Modelling: Meta-Modelling of 2D Hydrodynamic Model Using Artificial Intelligence Techniques (Support Vector Regression, Modified Linear Interpolation and Cellular Automata). FRMRC flood risk management consortium. FRMRC Research Report WP1.3.

Impacts of urban expansion on future flood damage: A case study in the River Meuse basin, Belgium

A. Mustafa
LEMA Research Group, ArGEnCo Department, University of Liege, Belgium

M. Bruwier
*Hydraulics in Environmental and Civil Engineering (HECE) Research Group,
ArGEnCo Department, University of Liege, Belgium*

J. Teller
LEMA Research Group, ArGEnCo Department, University of Liege, Belgium

P. Archambeau, S. Erpicum, M. Pirotton & B. Dewals
*Hydraulics in Environmental and Civil Engineering (HECE) Research Group,
ArGEnCo Department, University of Liege, Belgium*

ABSTRACT

Climate change and urban development are key factors influencing future flood damage. The main contribution of this paper is the evaluation of the sensitivity of future flood damage to different urban development scenarios. To do this, different urban expansion scenarios are proposed for the River Meuse basin in Belgium. Based on these scenarios, the impacts of urban development on flood damage are assessed. This study employs the multinomial logistic regression model that enables to visualize the consequence of different urban densities expansion. Cadastral datasets of years 1990, 2000 and 2010 are used to set four classes (non-urban, low-density, medium-density and high-density urban). Besides, several socio-economic, geographic and political driving forces dealing with urban development were operationalized to create maps of urbanization likelihood. These maps are then utilized to predict future urban expansion scenarios (high, medium and low expansion) for years 2020 and 2030 (Figure 1). The main factor controlling these scenarios is the future urban land demand. A validation of the urban expansion model showed that the model's outcomes allows to predict future urban expansion patterns with a relatively high explanatory power.

The estimation for flood damage of urban land-use was performed by overlaying inundation maps for the 100-year flood discharge plus 30% with the different urban expansion scenarios and by using a damage curve and specific prices. This paper considers only mobile and immobile prices for urban lands, which are associated to the most severe damages caused by floods along the River Meuse. While the urban expansion is higher for the low-density class, the flood damage increases more for the medium-density class where the flood induces higher damage.

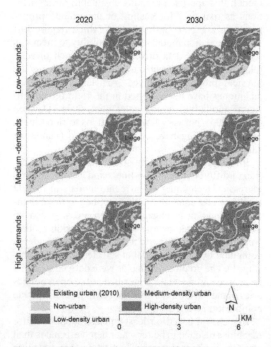

Figure 1. Urban scenarios of 2020 and 2030.

The increase of flood damage compared with 2010 are around 14–21% for 2020 and 23–38% for 2030. This work highlights that the use of several urban density classes to evaluate the flood damage enables to improve the computation since the price of each class can be better addressed.

Uncertainty assessment of river water levels on energy head loss through hydraulic control structures

P. Meert
Hydraulics Section, Department of Civil Engineering, KU Leuven, Leuven, Belgium

F. Pereira
Flanders Hydraulics Research, Antwerp, Belgium

P. Willems
Hydraulics Section, Department of Civil Engineering, KU Leuven, Leuven, Belgium
Department of Hydrology and Hydraulic Engineering, Vrije Universiteit Brussel, Brussels, Belgium

ABSTRACT

Physically based hydrodynamic models that solve the full de Saint-Venant equations provide a powerful to investigate the impact of man-made interventions like sluices, weirs and locks in the river system. Modelling these (controllable) hydraulic structures requires two types of information: firstly, the geometric properties of the structures and secondly, parameters such as head loss factors or discharge coefficients. The former type of parameters is relatively easy to obtain, whereas the latter are more difficult to assess. Therefore, default values are often used.

Good modelling practice and decision making processes are not only based on model results, but should also consider the uncertainties on these model results. In this paper, a sensitivity analysis was set up to investigate the uncertainties on the simulation results that arise from default parameter values. given that this is practically impossible with detailed hydrodynamic models, due to their long computation times, a lumped surrogate conceptual models was calibrated. This conceptual model is based on a concatenation of reservoir-type model elements and tries to mimic the results of the detailed model. Calibration of this conceptual model is done, based on simulation results of the hydrodynamic model. The explicit calculation scheme of these models allows for very short calculation times and makes them very useful for applications that require a large number of simulation or long term analyses.

In the presented conceptual modelling approach, the river network is schematized by a number of mutually interconnected storage reservoirs. Each reservoir has its own discretized mass balance equation to ensure continuity. The model set-up contains the exact same calculation scheme as the hydrodynamic model, to calculate the flow through different types of hydraulic structures. Furthermore, the model is also capable of accounting for changes in the water surface profile curve, due to changes in hydraulic structure discharge capacity.

The methodology is demonstrated by transforming a quasi-2D MIKE11 model into a conceptual model, for the case study of the river Dender in Belgium. This river has been struck by a number of severe flood events in the last two decades. The outdated beam weirs along the river, built in the 19th century, had an aggravating influence on the scale of these events. The Flemish government has therefore decided to renew the control structures in order to increase the discharge capacity and ensure their reliability.

The simulation results of the conceptual model show that the model is capable of reproducing the hydrodynamic model results with a significant reduction of calculation time (on average 80 times faster) and only minor accuracy loss (average RMSE lies below 5 cm for all water levels). These achievements are obtained for different boundary conditions and different parameter sets.

A Monte Carlo procedure with Latin Hypercube Sampling was used to perform sensitivity analyses of the considered parameters. The results of this analyses were then used to construct confidence intervals on the predicted river water levels. Results show that significant changes in water levels (up to 40 cm) can be observed. These changes are most distinct in the vicinity of the hydraulic structures and diminish for more upstream in the reach located water levels. Simulation results are most sensitive when the hydraulic structures operate in critical flow conditions. Extreme peak water levels, on the other hand, are not or only slightly affected by the considered parameter values.

Floods and cultural heritage: Risk assessment and management for the city of Florence, Italy

C. Arrighi & F. Castelli
Department of Civil and Environmental Engineering, University of Florence, Italy

B. Mazzanti
Arno River Basin Authority, Florence, Italy

ABSTRACT

Frequent flood events in the last decades emphasize the unique challenge for urban areas in managing natural hazards. Beyond the tangible potential losses to buildings, infrastructures and economic activities, historic cities have to face the threats on the cultural heritage. Bigio et al. (2011) show that Europe is on the top of the list for the number of World Heritage Cities (WHC) exposed to high and extreme flood risk. Increasing risk could lead to a loss of cultural assets that are felt by the population of immense importance for their contribution to human wellbeing (Mazzanti, 2003). Nevertheless, much more needs to be done in order to assess flood risk of cultural heritage. A quantification of the potential losses, direct and indirect, could accelerate the public debate and stimulate the adoption of protection measures. However, valuing the risk for the cultural heritage is a complex task since the value of an artwork is hardly monetizable. Here a case study of broad importance is presented, which is the art city of Florence (Italy), affected by a devastating flood in 1966. In Florence 176 buildings are officially classified as being part of the cultural heritage of the city (e.g. churches, museums, libraries etc.). For an estimated 100-year recurrence interval flood 46 of them may be affected by the inundation (Arrighi et al. 2016) The number increases to 126 for

Table 1. Damage categories for cultural buildings and artworks (left column) and assigned vulnerability level (right column).

Buildings	Vulnerability
Churches and religious buildings	high
Libraries, archives	low
Museums	medium
Noble palaces, theatres	medium
Artworks	Vulnerability
Paintings	0.60
Books, art prints	0.60
Sculptures	0.15
Goldsmith's art	0.10

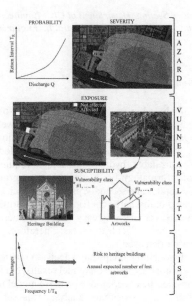

Figure 1. Sketch of the method for flood risk assessment to cultural heritage.

the 200-year event. Besides the cultural buildings, a huge number of ancient manuscripts, paintings, sculptures and art objects can be damaged. A preliminary risk assessment is presented for selected damage categories (Table 1) and possible risk mitigation strategies are discussed.

REFERENCES

Arrighi, C., Brugioni, M., Castelli, F., Franceschini, S. & Mazzanti, 2016. Flood risk assessment in art cities: the exemplary case of Florence, Italy. Journal of Flood Risk Management, in press.

Bigio A.G., Ochoa M.C., Amirtahmasebi R. & McWilliams K., 2011. Climate-resilient, climate-friendly world heritage cities, Urban Development Series. The World Bank.

Mazzanti, M. 2003. Valuing cultural heritage in a multi-attribute framework, microeconomic perspectives and policy implications. Journal of Socio-Economics, 32, 549–569.

Assessing the impact of climate change on extreme flows across Great Britain

L. Collet & L. Beevers
Heriot-Watt University, Edinburgh, UK

C. Prudhomme
CEH, Wallingford, UK

ABSTRACT

Floods are the most common and widely distributed natural risk to life and property worldwide, causing over £4.5B worth of damage to the UK since 2000. Climate projections are predicted to result in the increase of UK properties at risk from flooding. It thus becomes urgent to assess the possible impact of these changes on extreme high flows in particular, and evaluate the uncertainties related to these projections. This paper aims to assess the changes in extreme runoff for the 100-year return period event across Great Britain as a result of climate change and evaluate the associated uncertainties.

This study is based on the Future Flow database (Prudhomme et al., 2013) that provides daily river flow scenarios for 11 climate-change ensembles over 1961–2098. These ensembles reflect the uncertainties related to the HadRM3-PPE-UK model parameterization, as derived from the UKCP09 scenarios under the A1B emission scenario. This work analyses daily runoff for 281 gauging stations across Great Britain. The Generalized Extreme Value (GEV) distribution function (Coles, 2001) is automatically fitted over the baseline (1961–1990) and the 2080s horizon (2069–2098). The analysis evaluates the uncertainty related to the GEV distribution function and the climate model parameterization. Then it assesses return levels with combined uncertainties (see Fig. 1 for three contrasted gauging stations in Great Britain).

Results show that across Great Britain the GEV distribution generally computes spatially contrasted runoff estimates with the highest values on the west coast, and the lowest associated climate model and probabilistic distribution uncertainties. Similarly, the lowest estimates are found in south and southeast England, with the highest uncertainties of the country. From the baseline to the 2080s horizon, the GEV distribution shows increasing distribution function and climate model uncertainties. The climate model parameterization provides generally greater uncertainty than the GEV distribution function. The GEV distribution seems thus globally robustly defined for the 100-year return period estimation and quite sensitive to the input data it is fitted on.

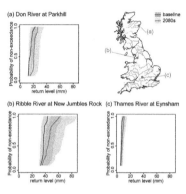

Figure 1. Combined uncertainties related to GEV distribution and climate model parameterization on three contrasted catchments: cumulated distribution function of the 11 100-year return period estimates and associated 95% confidence intervals over the baseline and the 2080s for (a) the Don River at Parkhill; (b) the Ribble River at New Jumbles Rock; (c) the Thames River at Eynsham.

Although this work accounts for various climate-change ensembles, such a statistical approach could be expanded to a probabilistic study where, with appropriate river flow database, numerous climatic scenarios would be accounted for, as derived from diverse climate models and representing diverse sources of uncertainties (Wilby, 2010). As the proportion of GEV distribution and climate model uncertainty is variable across the catchments, these uncertainties should systematically be combined and accounted for when assessing high return-period flows.

REFERENCES

Coles, S. 2001. *An Introduction to Statistical Modelling of Extreme Values*. Springer-Verlag, London.
Prudhomme, C., Haxton, T., Crooks, S., Jackson C., Barkwith, A., Williamson, J., Kelvin, J., Mackay, J., Wang, L., Young, A. & Watts, G. 2013. Future Flows Hydrology: an ensemble of daily river flow and monthly groundwater levels for use for climate change impact assessment across Great Britain. *Earth Syst. Sci. Data*, 5: 101–107.
Wilby, R. L. (2010). Evaluating climate model outputs for hydrological applications. *Hydrological Sciences Journal*, 55(7), 1090–1093.

Continuous hydrologic modeling of coastal plain watershed using HEC-HMS

M. Pourreza-Bilondi
Department of Water Engineering, University of Birjand, Birjand, Iran

S. Zahra Samadi
Department of Civil and Environmental Engineering, University of South Carolina, Columbia, South Carolina, USA

ABSTRACT

Accurate prediction of streamflow is an essential ingredient for both water quantity and quality management in the coastal plain watershed. Incorporating uncertainty forecast model helps enhance the reliability and credibility of the hydrological outputs and provides reliable information for effective decision making and water resources management. A consequence of this aggregation process is that most of the parameters in hydrology model cannot be inferred through direct observation in the field, but can only be meaningfully derived by calibration against an input-output record of the catchment response (Vrugt et al., 2008). GLUE has been used in many studies in hydrology (e.g. Vrugt et al., 2008; Vrugt et al., 2009). But its flexibility on The Hydrologic Modeling System (HEC-HMS; hereafter HMS) semi-distributed hydrology model was rarely studied before (Nourali et al., unpubl). The main aim of this paper is to construct MC based uncertainty assessment of HMS model to simulate streamflow over a highly complex and heterogeneous watershed.

This paper examines the effects of hydrologic model parameterization uncertainty or equifinality, where multiple unique hydrologic model parameter sets can result in adequate calibration metrics, on hydrologic modeling of the Waccamaw watershed (3116 km^2), a coastal plain watershed, in the southeastern US (SEUS). The well-known Hydrologic Engineering Center-Hydrologic Modelling System (HEC-HMS) is linked to the generalized likelihood uncertainty estimation (GLUE) and is calibrated for daily streamflow at two USGS gaging stations during 2003–2005 period. A lumped bucket-type model (the standard conceptual HEC-HMS's soil moisture accounting (SMA) approach) is applied in this study that represents a sub-basin with well-linked storage layers/buckets accounting for canopy interception, surface depression storage, infiltration, evapotranspiration, as well as soil water and groundwater percolation. The inverse error variance of GLUE model is then used to measure the closeness between model predictions and observations with 100,000 simulation runs using 37 parameters related to loss, transfer and baseflow. In this research, two indices were used to quantify the goodness of calibration/uncertainty performance: a P-factor, R-factor and Nash-Sutcliffe efficiency.

Analysis demonstrated that groundwater layer storage, percolation rate and storage coefficient are the most sensitive parameters in daily streamflow calibration. Further, 95 prediction uncertainty (95 PPU) bracketed most high and medium flows while slightly overestimated the magnitude and direction (increase or decrease) of low flow events. Since a large portion of the runoff-producing properties in the coastal plain watershed are groundwater dependent, hydrologic model parameterization may lead to statistically significant differences in streamflow prediction during different timescale.

REFERENCES

Nourali, M., Ghahraman, B., Pourreza-Bilondi, M. and Davari, K, unpubl. Choice of likelihood function for estimating uncertainty of HEC-HMS Flood Simulation Model Using Markov Chain Monte Carlo Algorithm, Journal of Hydrology, In review.

Vrugt, J.A., ter Braak, C.J.F., Clark, M.P., Hyman, J.M., Robinson, B.A., 2008. Treatment of input uncertainty in hydrologic modeling: Doing hydrology backward with Markov chain Monte Carlo simulation. Water Resour. Res. 44(12), W00B09.

Vrugt, J. A., Ter Braak, C. J., Gupta, H. V., & Robinson, B. A. 2009. Equifinality of formal (DREAM) and informal (GLUE) Bayesian approaches in hydrologic modeling? Stochastic Environmental Research and Risk Assessment, 23(7), 1011–1026.

A methodology to account for rainfall uncertainty at the event scale in fully distributed rainfall runoff models

I. Fraga, L. Cea, J. Puertas & M. Álvarez-Enjo
Environmental and Water Engineering Group, Departamento de Métodos Matemáticos y de Representación, Universidade da Coruña, Spain

S. Salsón & A. Petazzi
MeteoGalicia, Xunta de Galicia, Spain

ABSTRACT

The increasing importance of hydrological models as management and prediction tools has triggered the need of quantifying the uncertainty on their predictions. The uncertainty on the output of a hydrological model results from numerous factors. In this paper we present a new methodology to account for the spatial variability of rainfall uncertainty in distributed hydrological models.

Following the proposed methodology, a rainfall field is firstly interpolated from rain gauge data using the ordinary kriging (OK) technique. Then, a series of equally probable error fields are added to the OK field. In order to compute the error fields, first the standard deviation of the interpolation error at each spatial point of the OK field is estimated using the procedure described in Delrieu et al. (2014). Then, several random Gaussian fields are generated and the error on the rainfall estimation at each point is computed by multiplying the standard deviation of the error by the value of the Gaussian fields.

The presented methodology was validated using field data from 7 rainfall events registered in a 60 × 60 km region of the Northwest of Spain. Rainfall data from the meteorological agency MeteoGalicia (www.meteogalicia.es) rain gauge network was used to interpolate the rainfall in the study area and account for the prediction uncertainties. The predicted rainfall was compared to the rainfall measured by 3 additional rain gauges (installed within the CAPRI project, financially supported by the Spanish Ministry of Economy and Competitiveness). The results show a good performance of the proposed methodology regarding the estimation of uncertainties in the rainfall fields estimated by OK. The uncertainty bounds of the rainfall predictions covered more than 80% of the experimental data in all the events.

After validation of the presented methodology, we investigated the effects of the uncertainty in the OK fields on rainfall-runoff computations. In order to do so, we simulated an additional rain event in one of the

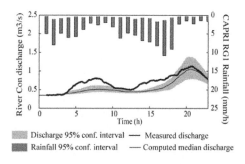

Figure 1. Caption of the computed discharges and predicted rainfall confidence bounds.

catchments located inside the study area. The numerical simulation was performed using the hydrological module of the software Iber (Bladé et al. 2014), which solves the 2D St. Venant equations to model surface runoff. A 2D Boussinesq model was implemented in Iber to account for groundwater flow, which is relevant in the study catchment.

The confidence bounds of the discharge at the catchment outlet pointed the significant effect of the rainfall uncertainty bounds on the computed hydrograph. During the second peak of the hydrograph, the bound amplitude of the predicted discharge is nearly 0.6 m^3/s, which represents around 50% of the peak discharge.

REFERENCES

Bladé, E., Cea, L., Corestein, G., Escolano, E., Puertas, J., Vázquez-Cendón, E., & Coll, J. 2014. Iber: herramienta de simulación numérica del flujo en ríos. *Revista Internacional de Métodos Numéricos para Cálculo y Diseño en Ingeniería* 30(1): 1–10.

Delrieu, G., Wijbrans, A., Boudevillain, B., Faure, D., Bonnifait, L. & Kirstetter, P.E. 2014. Geostatistical radar–raingauge merging: A novel method for the quantification of rain estimation accuracy. *Advances in Water Resources* 71: 110–124.

Estimating probability of dike failure by means of a Monte Carlo approach

R. Van Looveren, T. Goormans & J. Blanckaert
International Marine and Dredging Consultants nv, Antwerp, Belgium

Kristof Verelst & Patrik Peeters
Flanders Hydraulics Research, Antwerp, Belgium

ABSTRACT

The European Flood Directive 2007/60/EC requires Member States to assess the flood risk along their water courses and to take adequate measures to reduce this risk. Often flood risk maps are developed by only taking into account dike overtopping. However, floods can also be caused by dike failure. Not taking this into account leads to an underestimation of the flood risk. This paper presents a method to estimate the probability of dike section failure, taking into account several failure mechanisms. The method follows a Monte Carlo approach, in which all relevant parameters for the hydraulic loads as well as for the considered resisting forces vary according to specific distributions.

At the load side, the variation consists of a set of semi-probabilistic storms. The load at a dike segment is defined by the water level variation, the position of the phreatic line within the dike, and the wave conditions. Long-term measurements of all there variables, necessary for the statistical approach, are seldom available at the dike location. However, using (physically based) numerical models (or response curves derived from calculations with these models) and corresponding boundary conditions – i.e. known data of the variable at other locations – these variables can be estimated at the dike location. Statistical analysis of the boundary conditions results in knowledge of the probability of occurrence, which can be transferred to the dike location as well.

For the geotechnical variables often few data are available making it difficult to define the parameters of the probability distribution function (pdf). When data is available the pdf can be determined based on a statistical analysis. If not, often the case for geotechnical variables, the pdfs must be determined based on expert judgment and literature data. Since it is not possible to consider the full, continuous spectrum of sets of resisting parameter values, 'Latin Hypercube sampling' is used to compose the sets and determine their probability of occurrence.

Models are used to evaluate the limit state of each of the considered failure mechanisms: global instability, local instability, piping, erosion of the inner slope, and erosion of the outer slope. By evaluating the limit state for all combinations of sets of load and resistance, each with their own probability of occurrence, the probability of failure can be estimated (Monte Carlo approach). The global probability of failure is then calculated based on the failure probability of each mechanism.

To perform the necessary calculations the software tool BRES (Dutch for 'breach') was developed. The tool was developed in project WL_11_28, funded by Flanders Hydraulics Research. The method and tool have been successfully applied on various dike segments along navigable water courses in Flanders, both on tidal and non-tidal rivers. The paper describes two cases of application: a first application on the Meuse River, and a second one on the Scheldt River at Doel, in the framework of a to-be-constructed area with controlled reduced tide, situated next to the nuclear power plant of Doel.

The benefit of the approach compared to deterministic calculations is clear. Not only does the probabilistic approach enables the calculation of the global probability of failure of the dike section, also the probability related to the separate failure mechanisms is quantified. In addition, the full spectrum of load and resistance is considered, instead of only the most conservative ones. These elements provide a clearer view on those parts of the dike that require additional focus for possible reinforcement.

The methodology considers many phenomena in detail. However further improvements are possible. At the load side, for instance ship-induced waves and currents, or a varying wind profile, could be incorporated. At the resistance side, residual strength could be considered, additional dike revetment materials, or the phenomenon of water infiltration into the dike.

Developing a 100-point scoring system for quality assessment of ADCP streamflow measurements

H. Huang
Teledyne RD Instruments, Poway, CA, USA

ABSTRACT

The use of acoustic Doppler current profilers (often known as ADCPs) for streamflow measurements has increased significantly in the past ten years. The need to establish a framework with which to assess the quality of ADCP streamflow measurements becomes important. A quality assessment framework may consist of three components: (1) descriptive statistics, (2) uncertainty analysis, and (3) a scoring system. A scoring system assigns points to a number of quality indicators; the sum of all points is the score for a measurement. A scoring system provides a quantitative measure of the overall quality of a measurement.

Randall (2013) developed a scoring system which becomes a quality evaluation tool recommended by the Australia Water Information Standards Business Forum. A similar scoring system was developed and implemented by the New Zealand National Environmental Monitoring Standards. Although each of the scoring systems has its own merits, three observations can be made. First, both scoring systems do not use the discharge measurement uncertainty as a quality indicator. According to the International Organization for Standardization (ISO) document: "Guide to the Expression of Uncertainty in Measurement," uncertainty is the most important quality indicator for measurements. Second, in both scoring systems, a low score is associated with a high quality and a high score is associated with a low quality. This quality scaling is not easy to understand, particularly for ADCP users in some countries where a low score is often associated with a low quality and a high score is often associated with a high quality. Third, both scoring systems include some quality indicators that rely on operator's visual observation or subjective judgments, which may cause inconsistency in scoring among different operators.

This paper presents a 100-point scoring system for quality assessment of moving-boat ADCP streamflow measurements. The proposed scoring system employs a newly developed median-unbiased uncertainty estimator (Huang 2015) as a major quality indicator; it also considers four other quality indicators. The rating scale in the proposed scoring system is designed to be an analogy to the well-known 100-point wine rating scale. Table 1 shows the analogy between the two rating scales. The score point calculators for the quality indicators are presented. Results for ADCP streamflow measurements on five rivers are presented as application examples of the proposed scoring system.

Table 1. Analogy between the 100-point wine rating scale and the proposed 100-point rating scale for ADCP streamflow measurements.

Score	Wine (Newberry 2014)	ADCP streamflow measurements
96–100	Classic; a great wine; extraordinary	A^+: excellent
90–95	Outstanding; a wine of superior character and style	A: very good
80–89	Very good; a wine with special qualities	B: good
70–79	Mediocre; average; harmless	C: fair
60–69	Below average; deficient	D: meet minimum criterion
50–59 (or below 59)	Just downright unacceptable	F: fail to meet minimum criterion; unacceptable

REFERENCES

Huang, H. 2015. Optimal estimator for uncertainty-based measurement quality control. *Accreditation and Quality Assurance* 20(2): 97–106.

Newberry, B. 2014. What you need to know about the 100-point wine system. *The Savory*, available from: http://www.thesavory.com/drink/what-you-need-know-about-100-point-wine-system.html.

Randall, M. 2013. Application of acoustic Doppler's for determining velocity and discharge in open channels – *National Industry Guidelines* (Australia). Presented at the AIA Australia August 2013.

An alternate approach for assessing impacts of climate change on water resources: Combining hazard likelihood and catchment sensitivity

B. Grelier & G. Drogue
LOTERR, Université de Lorraine, Metz, France

M. Pirotton, P. Archambeau & B. Dewals
Hydraulic in Environmental and Civil Engineering (HECE), University of Liège, Liège, Belgium

ABSTRACT

This paper deals with the climate change impact on high flows and low flows of the Ourthe river at Tabreux (Meuse river catchment, Belgium).

Assessment studies of climate change impact on hydrology usually adopt top-down approaches that are largely driven by climate models outputs. For the present study, we aim to treat climate models outputs alternatively to the classical "predict and act" approach, by considering a large number of potential future climates including paleoclimate" reconstructions. By this way climate hazard assessment can be conducted through a likelihood approach (Prudhomme et al., 2010).

Plausible climate changes are estimated from atmospheric circulation indices of 13 CMIP5 climate models (RCP4.5 and 8.5) and "paleoclimate" records. A robust transfer function established at a mesoscale is used to link sea level pressure and geopotential at 500 hPa to precipitation and air temperature. The transfer function is extended back and ahead to estimate monthly multi-centennial surface climate variables series (1659–2100). Comparing the obtained series to the baseline precipitation and air temperature data allows producing more than 2000 monthly anomalies, which are used to transform the baseline daily climate data. As a result, a large number of scenarios are computed, documenting the likelihood of the climate change.

A hydrological rainfall-runoff (RR) model is provided with the perturbed daily series. To strengthen the method, the RR model is calibrated according to a classification of baseline sub-periods driven by the aridity index: a wet, an intermediate and a dry climate parameter set are defined and used to generate daily discharge series according to the aridity index values of the 30-year time slices of the target period.

Sensitivity to climate change of the Ourthe catchment at Tabreux is expressed as changes in quantiles of low flows and peak flows according to the baseline period: MAM7(5) and MAM7(10) as well as Q25 and Q100 are computed thanks to a Gamma statistical law (Ernst et al., 2010).

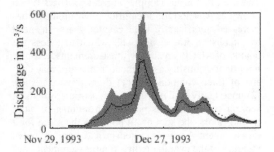

Figure 1. Observed and simulated hydrographs of the flood event of December 1993. The black dotted line represents the observed hydrograph. The black line represents the simulated hydrograph with baseline climate data. Grey lines correspond to the 2006 simulated hydrographs obtained with transformed climate data series.

Results show a predominant occurrence of drier climates in climate hazard in comparison to the baseline period, expressed by a great increase of air temperature, and a decrease of mean precipitation daily amounts. However, highest daily rainfall amounts are expected to increase. Low flows and peak flows are strongly variable according to climate hazard. In 60% (70%) of climate scenarios, Q25 (Q100) exceeds the baseline value up to a maximum value of 80% (90%). For 70% of the climate scenarios MAM7(5) and MAM7(10) values are under-exceeded the baseline value up to a maximum change of −75%. The figure 1 shows a variation of ±50% of the peak discharge of a significant flood event due to climate changes.

REFERENCES

Ernst, J., Dewals, B.J., Detrembleur, S., Archambeau, P., Er-picum, S. & Pirotton, M. 2010. Microscale flood risk analysis based on detailed 2D hydraulic modelling and high resolution geographic data. Natural Hazards, 55, 181–209.

Prudhomme C., Wilby R.L., Crooks S., Kay A.L. & Reynard N.S, 2010. Scenario-neutral approach to climate change impact studies: application to flood risk. *Journal of Hydrology*, 390, 198–209.

Efficient management of inland navigation reaches equipped with lift pumps in a climate change context

H. Nouasse, A. Doniec & E. Duviella
Mines Douai, IA, Douai, France
Université de Lille, France

K. Chuquet
Voies Navigable de France, Service de la navigation du Nord, Lille, France

ABSTRACT

A historic agreement to combat climate change was adopted by 195 countries, the 12th of December 2015 in Paris. The countries will *pursue efforts to* limit the temperature increase to 1.5°C from 2100. Even if this objective is reached, the Intergovernmental Panel on Climate Change (*IPCC*) expects a modification of the rain distribution over seasons and areas by the end of the century. RCP scenarios (*Representative Concentration Pathways*) on which future forecasts on temperature and rain have been generated and studies on the frequency and intensity of future flood and drought periods have been achieved and are available in the literature. Recent studies address the climate and Human impact on future hydrological drought across the world. The conclusions are that hydrographical networks will be impacted by drought events, particularly the inland navigation networks. The available water resource used to supply the inland navigation networks will be insufficient during drought periods.

The main objective of the inland navigation network managers is to gather all the conditions guaranteeing the navigation accommodation. To achieve this aim, it is necessary to control the water volumes in each parts of the networks. It consists in supplying or emptying the network and then dispatching and balancing the water resource in all the network taking into account the navigation demand and the climate hazards like flood and drought periods. These extreme climate events are expected more frequent and stronger in the future. For example, in North of France, the results of the *Explore*2070 project[1] expect a decrease between 15 and 45% of the available water resource from 2050–2070. Hence, the design of an efficient water management for inland navigation network becomes a challenging objective.

Hence, adaptation measures allowing the system to function have to be proposed. To achieve this objective, it is first necessary to determine the resilience of the inland navigation networks against the drought periods, *i.e.* their capacity to allow the navigation. Secondly, adaptive management strategies have to be designed to optimizing the water resource management. Finally, it is often necessary to propose structural adaptations of the networks in order to improve their resilience.

In this paper, we propose an efficient management strategy of inland navigation networks equipped with lift pumps. It aims at guaranteeing the navigation conditions by minimizing the pumping cost. It is based on based on a quadratic optimization problem. To design this approach, an integrated model of inland navigation networks is firstly proposed. It allows reproducing the dynamics of each navigation reaches that compose the networks. All possible interactions between the network and its environment are taken into account. The dynamical variables are the water volumes on several hours. This integrated model is then transformed with a flow graph formalism. The proposed flow graph aims at designing the quadratic optimization problem.

The proposed efficient management strategy is applied on an inland navigation networks composed of 2 reaches. It is shown that in a climate change context drought events could disturb the navigation particularly when less water volumes are available and structural management of the network is required. Considering the equipment of lift pumps, the efficiency of the proposed management strategy to optimize the water volume management at low cost is highlighted.

ACKNOWLEDGEMENT

This work is a contribution to the GEPET-Eau project (http://gepeteau.wordpress.com/enversion/) which is granted by the French ministry MEDDE – GICC, the french institution ORNERC and DGITM.

---

[1] http://www.developpement-durable.gouv.fr/Evaluation-des-strategies-d.html

Evaluation of changes storm precipitations during century for the modeling of floods

V.V. Ilinich & T.D. Larina
Department of Meteorology, Hydrology and Water Flow Regulation,
Institute of Environmental Engineering of Russian State Agrarian University, Moscow, Russia

ABSTRACT

Every dam of water reservoir must be projected in respect to catastrophic flood of normative probability exceeding relative to maximum of water discharge and volume of flood. Some water reservoirs are located on the small river basin. Usually we not have ranks observations for the runoff on small rivers. Consequently there are different models and formulas for determination of flood characteristics in dependence on storm rainfall. Different probabilistic quintiles of maximum daily precipitation are used in the models and formulas. Some researchers have written hypothesis about increase of maximum daily precipitation (Groisman 2005, Dore 2005, Wilks 1999). However we must have very long time series of observation relative to daily precipitation for the checking of the hypothesis. The research was dedicated to the checking of such hypothesis. Several problems were solved in the work: selection of meteorological stations which have very long time series of observation the daily rainfall; the forming statistical ranks for different discrete values of precipitation and assessment their trends; determination of the main statistical characteristics relative to the formed ranks and evaluation of their changes in the time; evaluation of the changes in the time of frequencies of appearance dangerous storm rainfalls; evaluation of the changes in the time of probabilistic quantiles of maximum daily precipitation in the frames of concrete distributions of random values. The results have showed that maximums of daily precipitations are increasing during the time on majority of researched meteorological stations.

All research aspects prove that there is increasing of the daily dangerous extreme precipitations on majority of meteorological stations of the west part of the North Caucasus territory.

Question about the correct mathematical assessment of the daily dangerous extreme precipitations in respect to their probabilities remains open.

Statistical rank of extreme precipitation and maximum river discharges for the last 30 years can have quantiles which exceed significantly quantiles of full statistical rank. This is necessary to take to consideration in the hydrological calculations.

It is justified to use some non-standard methods for asymmetry evaluation in respect to meteorological and hydrological time series observations in frames of concrete distributions of random values.

In the presence of long meteorological observations, there is no need to build analytical curves of probability for definitions of quantiles of probabilities of $P = 5\%$ and $P = 10\%$ depending on the parameters Cv and Cs in the cases of presence of long meteorological observations (over 30 years). Such curves are required for assessment of dangerous phenomena for estimated quantiles of the distribution ($P = 0.1\%$; $P = 0.5\%$; $P = 1\%$).

REFERENCES

Dore, MHI. 2005. Climate change and changes in global precipitation patterns: what do we know. Environ. In. 31(8): 1167–1181.

Dmowska, Ed.R. & Hartman D. & Rossby H.T. 2011. Statistical methods in the Atmospheric Sciences. Inter. Geoph. Series Vol. 1. Oxford, OX51GB, UK: 668.

Groisman, P.Y. 2005. Trends in intense precipitation in the climate record. J. Clim.: 18, 1326–1350.

Ilinich, V.V. 2010. Assessment of skew in the frames of three-parametric gamma distribution. J. Prirodoobustroystvo, No. 5 (In Russian): 71–75.

Ilinich, V.V. 2014. Evaluation of asymmetry for ranks of extreme hydrological values. Proceedings of Conference: "21 century: fundamental science and technology III", Vol. 3, North Charleston, SC, USA 29406: 10–13.

Kristoforov, A.V. 1988. Theory of probability and mathematical statistics. Moscow: 128 (In Russian).

Krtskiy, S.N. & Menkel, M.F. 1981. Hydrological basis of river flow management. Moscow: 255 (In Russian).

Wilks, D. 1999. Inter annual variability and extreme-value characteristics of several stochastic daily precipitation models. Agril. AndForest Met. 93(3):153–169.

Main impacts of climate change on seaport construction and operation

B. Koppe
Institute for Hydraulic Engineering, City University of Applied Sciences Bremen, Bremen, Germany

ABSTRACT

Due to their location at the intersection between sea and land, marine facilities are most vulnerable to changes of all water related parameters like mean relative sea level, storm water levels, wind waves and swell, tidal regime, sedimentation rates, waterborne immigration of species, water salinity and acidity. Furthermore, seaports can also be affected directly by temperature, precipitation and wind changes with respect to e.g. cooling system energy demands, terminal pavement durability, storm-water drainage system capacity, empty container storage heights etc.

It is difficult to make general statements on the vulnerability of seaports to climate change. In some cases, already today climate change related problems in seaports are obvious, like changing properties of permafrost soils and related foundation problems in high latitudes. Other climate change impacts will influence port planning and operation only at a later stage, because in short and mid-term only minor effects are effective or the vulnerability of the system 'seaport asset' to the respective climate trend is only marginal. Additionally, ports located at the open sea have to bear other loads than estuary or lock-separated ports and the sensitivity of cargo handling with respect to wind and wave loads differs with cargo type.

Restrictions for assessment and management of climate change challenges to seaports are manifold as it is still difficult to reliably predict climate change effects and to deal with the long time spans climate change refers to. Some of the main restrictions are differences in planning horizons, lack of relevant information, treatment of uncertainty, and financial aspects. In Table 1 differences in planning horizons and lifetime of port facilities are shown. Climate change projections refer to time spans of decades to centuries because the climate variability makes it difficult to generate any short-term projections of less than 25 years. In contrast, today's port master plans are laid out for a time span of at the most 15 to 25 years, not least because industrial planning horizons seldom exceed 15 years. In contrast, the minimum design life of most port related infrastructure is around 30 to 60 years, whereas design lives of 100 or more years are assigned to port protection structures.

As a basis for climate change adapted processes in port planning and operation a matrix containing possible climate change impacts and possibly affected port assets is presented. Furthermore, steps of a vulnerability analysis of seaports against climate change effects are described and the sensitivity of specific port assets to climate change effects as well as possible adaptation are exemplified. Combining the analysis of vulnerabilities of specific seaport assets to climate change and the development of appropriate active or reactive adaptation measures an effective climate proof planning and operation procedure can be established in ports and harbours.

Table 1. Differences in planning horizons and lifetime of port facilities.

Port planning / Port super- and infra-structure	Planning horizon / Lifetime in years									
	10	20	30	40	50	60	70	80	90	100
Most common planning horizons of port master plans	■									
Far-reaching planning horizons of port master plans	■	■	■							
Terminal and port superstructure	■	■	■							
Pavements	■	■	■							
Berth structures serving special industries	■	■	■	■						
Open piers (open piled structures)	■	■	■	■	■					
Dry docks	■	■	■	■	■	■				
Shore protection works	■	■	■	■	■	■	■			
Quay walls	■	■	■	■	■	■	■	■		
Flood protection				■	■	■	■	■	■	■
Breakwaters			■	■	■	■	■	■	■	■

Special session: Innovative solutions for adaptation of European hydropower systems in view of climate and market changes

Experimental assessment of head losses through elliptical and sharp-edged orifices

N.J. Adam, G. De Cesare & A.J. Schleiss
Laboratory of Hydraulic Constructions, École Polytechnique Fédérale de Lausanne, Lausanne, Switzerland

ABSTRACT

Most of Swiss storage power plants are high head power plants exploiting water heads higher than 200 m. They can provide large amount of energy in a short lapse of time. In a high competitive market, it is interesting to increase their flexibility. Following a refurbishment with discharge increase, surge tank geometry should be checked to insure the safety. Due to several physical limitations, e.g. topographical, the placement of an orifice at the entrance of a surge tank is the most economical way to adapt the surge tank.

Surge tank orifices allow to reduce extreme water levels in surge tanks. Unidimensional analysis are performed to determinate head losses [1], which can be symmetric or asymmetric. However, it is sometimes difficult to determine the corresponding orifice shape as head losses are quite sensitive to all geometrical parameters.

In this study, the experimental set-up is a straight PVC pipe with a diameter $D = 0216$ m and a total length of 4 m. Orifices are placed at the middle of the pipe and water flows in both directions. Two orifice geometries are tested: Standard [2] and elliptical orifice geometry (Figure 1). Each elliptical orifice, whose geometrical parameters are defined by the least square method, are related to each sharp-edged orifice.

For all orifices, head losses are evaluated for ten different discharges from Reynolds number in the main pipe from $6 \cdot 10^4$ to $18 \cdot 10^4$. Hydraulic head are assessed along the pipe axis to evaluate head losses produced by the orifice in both directions.

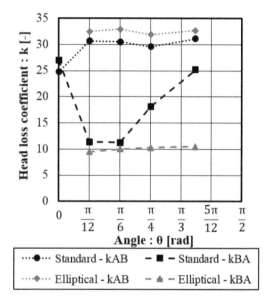

Figure 2. Head loss coefficients for sharp-edged and elliptical geometries.

The influence of the sharp-edged angle θ on head losses are evaluated and shown in Figure 2. Figure 2 shows that sharp-edged orifices allow to introduce different values of asymmetry with angle variation. This angle introduction increases by around 20% in one direction and decreases by more than 50% in the other direction. However, for elliptical orifices, the elliptical shape does not significantly influence head losses in both flow directions.

REFERENCES

Alligne, S., Rodic, P., Arpe, J., Mlanick, J. and Nicolet, C., 2014. Determination of surge tank diaphragm head losses by CFD simulations. Environmental Advances in Hydroinformatics, 325–336.

Internation Standard, 2003. ISO 5167-2: Measurment of fluid flow by means of pressure differential devices inserted in circular cross-sections conduits running full-part 2: Orifice plates.

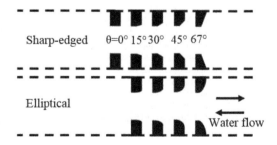

Figure 1. All orifice geometries tested in the present study.

Integrated assessment of underground pumped-storage facilities using existing coal mine infrastructure

R. Alvarado Montero, T. Wortberg, J. Binias & A. Niemann
Institute of Hydraulic Engineering and Water Resources Management, University of Duisburg-Essen, Germany

ABSTRACT

The network of tunnels in the Prosper-Haniel mine located in the Ruhr region in Germany has been analyzed as a possible lower storage for the development of a pumped-storage project. These tunnels can hold approximately 600,000 m^3 at depths that range between 600 and 1000 m. The conditions make it a suitable location to store large amounts of energy (up to 1700 MW-h) for short periods of time, which contributes to balance the regional energy grid. Despite the potential of this existing network of tunnels, the future under-ground water levels will compromise its use, and therefore alternative options to excavate a new storage structure have been proposed. This paper describes the most relevant hydraulic aspects for developing the project, considering construction, geotechnical, geological, and energy market restrictions. An overview of the most relevant conclusions reached during the assessment of Prosper-Haniel mine are presented and a set of recommendations is provided for future assessments of underground pumped-storage facilities.

Some of the most relevant characteristics of each of the options analyzed during this project are summarized in Table 1. Even though the cost per kW for the options of level 6 is below 1000 Euros, the potential development of pumped-storage in this level is compromise by the future underground water levels.

The ring structure, which consists of excavating 15.5 km of new tunnels, takes advantage of existing shafts infrastructure. However, it is still an expensive alternative with costs/kw above 2750 Euros. It offers about 820 MWh for each production cycle in the system with total power capacity of 200 MW.

The results presented here are part of a first phase of this project, which looked into the technical and economic feasibility of using the existing mine infrastructure for the potential development of a pumped-storage facility. The analysis was a first step and by no means was expected to conclude with a final design of the project. The infrastructure in Prosper-Haniel mine, however, has been selected as one of the most convenient locations to develop an energy-storage project at an innovative location. The second phase of this project is projected in 2016 and it is expected to include a refinement of the results from the first phase, together with a more detailed design. For further information please visit: www.upsw.de/

Table 1. Main characteristics for each of the analyzed options in Prosper-Haniel (PH) mine. Option 6a refers to the currently open tunnels in the network (Feasibility level).

Characteristic	PH New Ring	PH[1] Level 6a	PH[2] Level 6b
Total cost (M. Euro)	560	260	300
Cost/kW (Euro)	2750	760	870
Reservoir length (km)	15.5	22.2	28.8
Net water head (m)	560	985	985
Reservoir volume (m^3)	600,000	555,000	720,000
Power (MW)	200	350	350
Energy/cycle/MWh	820	1340	1740
Cost/MWh (Euro)	680	194	172

[1] level 6a corresponds to the network of existing open tunnels.
[2] level 6b includes partly closed tunnels in addition to the existing network of open tunnels.

REFERENCES

Alvarado Montero, R., Niemann, A., Schwanenberg, D. 2013. Concepts for Pumped-Storage Hydroelectricity using Underground Coal Mines. *Proceedings of 2013 IAHR Congress*, Tsinghua University Press, Beijing.

Alvarado Montero, R., Niemann, A., Wortberg, T. 2015. Underground Pumped-Storage Hydroelectricity using Existing Coal Mining Infrastructure. *Proceedings of the 36th IAHR World Congress*, The Hague, the Netherlands.

German Advisory Council on the Environment (SRU). 2011. *Pathways towards a 100% renewable electricity system.* Ed. Erich-Schmidt GmbH, Berlin.

Niemann, A.; Perau, E.; Schreiber K.; Koch, H.-J. (2014): Chancen und Risiken untertägiger Pumpspeicherwerke in Steinkohlebergwerken im Ruhrrevier, Zeitschrift Wasserwirtschaft 1/2 2014, S. 66–69; ISSN: 0043-0978.

Feasibility assessment of micro-hydropower for energy recovery in the water supply network of the city of Fribourg

I. Samora
Laboratory of Hydraulic Constructions, École Polytechnique Fédérale de Lausanne, Lausanne, Switzerland
Civil Engineering Research and Innovation for Sustainability, Instituto Superior Técnico, Lisbon, Portugal

P. Manso, M.J. Franca & A.J. Schleiss
Laboratory of Hydraulic Constructions, École Polytechnique Fédérale de Lausanne, Lausanne, Switzerland

H.M. Ramos
Civil Engineering Research and Innovation for Sustainability, Instituto Superior Técnico, Lisbon, Portugal

ABSTRACT

The present concerns with environment and energy efficiency contribute for a growing new interest on micro-hydropower. One of the most promising applications of micro-hydropower lies within water supply systems. These are systems which are pressurized and where the pressure control is important to avoid water losses and pipe damage (Carravetta et al. 2012). The use of turbines instead of pressure reduction valves for the excess pressure dissipation allows to recover part of this energy (Ramos et al. 2010, McNabola et al. 2014).

However, there is still a lack of technologies and specific solutions for energy recovery in these conditions, in particular for meshed networks. In the present work, a feasibility study is carried for the installation of micro-turbines within the city of Fribourg, Switzerland.

A possible hydropower arrangement based on a new and recently tested turbine, the 5BTP, is proposed. Up to four turbines can be installed within buried chambers, as presented in Figure 1.

To identify the optimal placement of the chambers, an algorithm was developed. This algorithm is an upgraded version of a previously developed work (Samora et al. 2016). The optimization was carried out considering the net present value after 20 years of operation ($NPV_{20years}$) – Table 1.

For the unit prices and sell tariffs assumed, the arrangement is economically feasible for the installation of up to four turbines in the case study. It came to

Table 1. Results for the installation of four turbines.

Pipe ID	E (MWh/year)	NPV$_{20years}$ (kCHF)		
		r: 4%	6%	8%
(2730, 2730)	121	513	428	362
(2730, 2730, 2730)	128	546	456	386
(2091, 2730, 2730, 2730)	144	573	473	396

no surprise that the pipe chosen in all solutions belongs to the path with the highest discharge flows and where pressure reduction valves are already installed. However, the best configuration for four turbines imply placing the turbines in multiple pipes. The reduction of the flow discharge due to the insertion of a first chamber and corresponding head-loss justifies the interest of building a second chamber. Finally, the effect of the installation of a by-pass for construction and maintenance was analyzed, which has an important weight in the investment, representing 30% of the total costs.

REFERENCES

Carravetta, A., Giuseppe, G., Fecarotta, O. & Ramos, H. 2012. Energy production in water distribution networks: A PAT design strategy. *Water Resources Management*, 26(13): 3947-3959.

McNabola, A., Coughlan, P., Corcoran, L., Power, C., Williams, A. P., Harris, I., Gallagher, J. & Styles, D. 2014. Energy recovery in the water industry using micro-hydropower: an opportunity to improve sustainability. *Water Policy*, 16:168–183.

Ramos, H., Borga, A., & Simão, M. 2009. New design for low-power energy production in water pipe systems. *Water Science and Engineering*, 2(4):69–84

Samora, I., Franca, M. J., Schleiss, A. J. & Ramos, H. M., 2016. Simulated annealing in optimization of energy production in a water supply network. *Water Resources Management*, 30(4): 1533–1547.

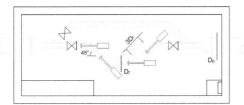

Figure 1. Schematic lay-out of a chamber equipped with four turbines.

Banki-Michell micro-turbines for energy production in water distribution networks

V. Sammartano & P. Filianoti
Department of Civil, Energy, Environmental and Materials Engineering, Università Mediterranea di Reggio Calabria, Reggio Calabria, Italy

G. Morreale
WECONS, Palermo, Italy

M. Sinagra & T. Tucciarelli
Department of Civil, Environmental, Aerospace, Material Engineering, Università degli Studi di Palermo, Palermo, Italy

ABSTRACT

Banki-Michell turbines combine simplicity with efficiency and represent a possible alternative to the use of PATs (Pump As Turbines) for hydropower energy production at the end of aqueducts delivering water to urban tanks (Sinagra et al. 2014). The traditional outlet of this type of turbine is at atmospheric pressure and this precludes its allocation along aqueducts or inside Water Distribution Networks (WDNs). The use of in-line turbines could allow the selection of these sites, where a pressure reduction would not affect discharge regulation, but rather limit the pressure downstream of the turbine and reduce water losses. Starting from the design criteria previously developed for Banki-Michell turbines (Sammartano et al. 2015), a new Banki-Michell turbine aimed to overcome this limitation is proposed. A detailed analysis is carried out to investigate about the variation of both the efficiency and the total head dissipation along with the change of the impeller rotational velocity, assuming the electric regulation of the turbine to be available. A scheme of the new turbine is reported in Figure 1. It is composed by four main parts: the convergent pipe, the nozzle, the rotating impeller and the outlet pipe.

An appropriate design procedure was carried out taking into account the following project input data: a specific energy drop ΔH of 10 m, with an upstream head, $H_m = 35$ m and a downstream head $H_V = 25$ m; a water discharge $Q = 0.03$ m^3/s; a rotational velocity of the impeller $\omega = 750$ rpm. The turbine geometry

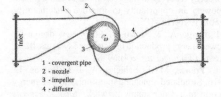

Figure 1. Scheme of the Banki-Michell turbine.

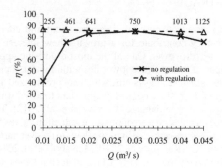

Figure 2. The efficiency curves of the turbine in the two considered configurations: with and without regulation system.

was optimized performing a series of CFD simulations solving the RANS equations and selecting the SST turbulence model. The efficiency of the turbine was numerically tested taking into account different values of the flow rate (0.010–0.045 m^3/s) and maintaining a constant impeller rotational velocity ω. Other simulations were carried out in order to investigate about the benefit of controlling of the rotational velocity, in order to keep constant the velocity ratio of the BEP point, $V_r = 2$. The efficiency curves are reported in Figure 2. The results are quite encouraging, especially if an electric regulation system is used to control the impeller rotational velocity: the efficiency is always greater than the 84.3% with a peak value of 86.8%.

REFERENCES

Sammartano, V., Aricò, C., Sinagra, M., & Tucciarelli, T. 2015. Cross-Flow Turbine Design for Energy Production and Discharge Regulation. Journal of Hydraulic Engineering, 141(3), March 2015.

Sinagra, M., Sammartano, V., Aricò, C., Collura, A., & Tuccia-relli, T. 2014. Cross-Flow turbine design for variable oper-ating conditions. Procedia Engineering, 70(2014), 1539–1548.

How groundwater interactions can influence UPSH (Underground Pumping Storage Hydroelectricity) operations

S. Bodeux, E. Pujades, P. Orban, S. Brouyère & A. Dassargues
University of Liège, Hydrogeology & Environmental Geology, Aquapole, ArGEnCo Department, Engineering Faculty, Belgium

ABSTRACT

In the current energy grid, renewable energy has an increasing role to play. However, their intermittence cannot afford to regulate the produced electricity according to the irregular demand (Evans *et al.*, 2012). Pumped Storage Hydroelectricity (PSH) is a well-known efficient technology to store and release electricity according to the demand needs but appropriate potential new sites are getting scarce (Steffen, 2012). An innovative alternative consists in using abandoned mines as lower reservoir of an Underground Pumping Storage Hydroelectricity (UPSH) plant (Fig. 1).

In such configuration, large amount of water will be pumped or injected in underground cavities, creating subsequently head oscillations in the surrounding aquifers. Consequently, this seepage occurring between the considered cavity and the varying groundwater heads in the surrounding geological medium may influence the efficiency of the UPSH plant but also the magnitude of the potential impacts on the groundwater resources.

A hybrid 3D finite element mixing cell method (Brouyère *et al.*, 2009) is used to simulate numerically the use of a representative UPSH cavity and calculate the induced changes in groundwater heads in the surrounding geological medium. Different scenarios are computed varying parameter values (hydrogeological and lower reservoir characteristics), boundary conditions, and pumping/injection time-sequences.

By analyzing the computed piezometric heads at different distances from the underground reservoir, the magnitude of the aquifer response to pumping storage operations is assessed. The most expected and noticeable effect is the oscillation of groundwater levels. The existence a mean pseudo/dynamic steady-state and the required time to reach it are also determined. The head difference and its time evolution between the cavity and the surrounding medium is triggering the leakage of groundwater into the cavity or the contrary. The resulting effects on the UPSH plant efficiency can be estimated.

Combining these outcomes, some feasibility criteria of this type of projects are identified. Going into

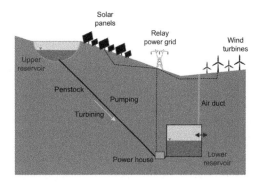

Figure 1. Schematic representation of an UPSH.

practice, further models should include more in details local and specific geometrical and hydrogeological data of the considered old mine cavities used as lower reservoir. This kind of modelling approach can be used as a first approach for determining how the aquifer will response to short and long term changes in UPSH pumping/injection schemes.

REFERENCES

Bodeux S., Pujades E., Orban Ph. and Dassargues A., 2016, Mine as lower reservoir of an UPSH: groundwater impacts and feasibility, EGU2016, Vienna, April 17–22.

Brouyère S., Orban P., Wildemeersch S., Couturier J., Gardin, N. and A. Dassargues, 2009. The Hybrid Finite Element Mixing Cell method: A new flexible method for modeling mine ground water problems. *Mine Water Environment*, 28(2): 102–114.

Evans A., Strezov V. and Evans T.J., 2012. Assessment of utility energy storage options for increased renewable energy penetration. *Renewable and Sustainable Energy Reviews*, 16(6): 4141–4147.

Pujades E., Willems Th., Bodeux S. Orban Ph., Dassargues A. 2016, Underground Pumped Storage Hydroelectricity (UPSH) using abandoned works (deep mines or open pits): impacts on groundwater flow, *Hydrogeology Journal*, in press.

Steffen B., 2012. Prospects for pumped-hydro storage in Germany. *Energy Policy*, 45: 420–429.

Integration of hydropower plant within an existing weir – "A hidden treasure"

M. Marence & J.S. Ingabire
UNESCO-IHE, Delft, The Netherlands

B. Taks
DEC Dommelstroom U.A. Sint-Michielsgestel, The Netherlands

ABSTRACT

Most of the best feasible hydropower sites in Europe have been developed and only half of technically feasible potential is still available for development. Construction of each new hydropower plant in Europe is loaded by very strong opposition and restrictive existing environmental and administrative procedures. These restrictions caused nearly full stagnancy in erection of new hydropower plants with capacity above 10 MW. Small scale hydropower potential is widely recognizable and development opportunities are significant, but implementation struggles also on restrictions and negative image. High construction costs and low energy prices in Europe, at the moment, additionally discourage new developers. Installation of new power plants in the existing water structures without hindering their primary functions could become an attractive solution, saving construction costs and minimized environmental impacts and negative public image.

In this regards, the study of use of an existing weir structure developed for water management and navigation purposes, has been developed. The weir structure on Dommel River, in Sint Michielsgestel, province of North Brabant, the Netherlands has been constructed in 1970 is selected and analyzed in this study.

The constant denivelation and discharge in the weir site give an ideal condition for installation of a small hydropower plant. The private investor started the licensing process. During the licensing process, UNESCO-IHE has prepared a technical study checking and optimizing the site (Ingabire, 2014).

The study included all of the influenced parameters and tries to find an optimal solution and the optimal setup for the planned power plant. Multi-criteria analysis has been performed for selecting optimal power plant setup and also optimal turbine type.

The solution with a power plant merged in one of the weir fields is suggested and selected. Different types of low head turbines have been analyzed with special consideration on construction opportunities, efficiency, flexibility and construction costs. Based on given boundaries and restrictions, the solution with a screw turbine has been seen as the most promising option. The turbine can be installed in one of the weir fields with minimal reconstruction works. The lower efficiency and therefore power output compared to some other turbine types could be compensated with lower turbine construction costs and simpler construction and large turbine flexibility in head and especially discharge variation.

A short overview of the hydropower situation in the Netherlands including hydro potential, possible solutions and trends, tariffs and legal regulations, and environmental and social constraints is also given.

REFERENCES

Bölli, M. & Feibel, H. (2015): Low hanging fruit: hydropower in existing water infrastructure. *ESI Africa*, (1): 76–77.
ICOLD (2007): Dams & The World's water. International Commission on Large Dams, www.icold-cigb.org.
Ingabire, J.S. (2014): Integration of hydropower plant within an existing weir structure – a "hidden treasure", Case of Sint Michielsgestel weir, North Brabant, the Netherlands. MSc. Thesis UNESCO-IHE, WSE HERBD 14.05.
Marence, M. (in press). Western Europe review, In *World small Hydropower development Report 2016*, UNIDO and IC-SHP.
Paish, O. (2002): Small hydropower: technology and current status. Renewable and sustainable energy reviews 6(6): 537–556.
US DOE (2012): An assessment of Energy potential at Non-Powered Dams in the United States. The United States Department of Energy.

Special session: Buoyancy-driven flows

Numerical simulation of periodically forced convective currents in aquatic canopies

M. Tsakiri & P. Prinos
Department of Civil Engineering, Aristotle University of Thessaloniki, Thessaloniki, Greece

ABSTRACT

In the present study, periodically forced convective currents between open water and emergent vegetation are investigated numerically. These currents are produced due to differential heating and cooling between the two regions, during a typical diurnal cycle. In particular, during the daytime, the vegetation prevents the solar radiation from entering into the water body and thus, the water temperature in the open area is higher. During the night-time, the vegetation reduces the temperature losses and thus, the open water cools faster.

The simulation is referred to lock-exchange flow in a tank partially covered by cylinders which represent the emergent vegetation (Fig. 1). The surface of the open water is imposed to periodical thermal forcing. The unsteady two-dimensional Volume-averaged Navier-Stokes equations together with the energy equation are solved using the Fluent CFD code. The Boussinesq approximation is used for taking into account the density difference due to temperature difference. The radiation absorption of the water during the daytime and the heat loss of the water body during the night-time are simulated by an internal source which varies with time periodically and also with the water depth. This additional source is added in the energy equation. The vegetation effects are taken into account through additional terms in the momentum equations. The term of the resistance drag due to vegetation is based on the porous media theory and includes both the linear and the non-linear term.

The model tank is 1.2 m long consists of a vegetated and a non-vegetated region. Three cases with different vegetation porosity φ and one case without vegetation ($\varphi = 1.0; 0.95; 0.85; 0.75$) are examined for investigating the effect of vegetation porosity in the convective currents. Initially the tank contains water of constant density and temperature. At $t = 0$ s, a periodical internal thermal force is imposed on the open water. The period of thermal force is equal to $P = 700$ s and the maximum surface intensity is 300 W/m^2. The simulation lasts 10 full thermal forcing cycles for minimizing the effect of the start-up flow.

For $\varphi = 1$ (no vegetation), the horizontal velocity contours indicate three distinct regions of circulation at $t = 9$, 9.75 and 10P due to cooling phase. At $t = 9.25$ and 9.5P, these circulations have the opposite direction. For $\varphi = 0.85$, the velocity within the vegetation is almost zero. At $t = 9$ and 10P, two distinct regions of circulation are observed in the open region, while at $t = 0.5P$ these circulations have the opposite direction. At $t = 0.25P$ one distinct region of circulation is observed in the open area due to resistance drag and at $t = 9.75P$ this circulation has the opposite direction.

From the horizontal velocity profiles at the lock gate, it is shown that, the velocities of both the surface and the bottom current, during the heating phase, are constant for $\varphi = 1$. During the cooling phase, the currents return back with the same velocity magnitude. For $\varphi = 0.85$, the current velocity is not constant during the heating or the cooling phase and it is lower than that for $\varphi = 1$ due to the resistance drag.

The flow response is out of phase with the thermal forcing and there is a lag time which is increased with decreasing vegetation porosity. Moreover the flow rate at the lock gate is decreased with decreasing vegetation porosity. Finally the volume-averaged temperature in the open region is increased with decreasing vegetation porosity in the vegetated region, while that in the vegetated region is decreased.

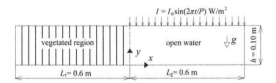

Figure 1. Model geometry.

Experiments on the impact of snow avalanches into water

G. Zitti
D.I.C.E.A. – Università Politecnica delle Marche, Ancona, Italy

C. Ancey
F.E.N.A.C. – Ecole Polytechnique Fèdèrale de Lausanne, Lausanne, Switzerland

M. Postacchini & M. Brocchini
D.I.C.E.A. – Università Politecnica delle Marche, Ancona, Italy

ABSTRACT

The number of anthropized reservoirs threatened by snow avalanches is steadily increasing and impulse waves caused by avalanche impact are becoming a considerable risk for such basins. The dynamics of the impact of a snow avalanche into a water body is studied through laboratory experiments, where a granular material, with solid density slightly lower than that of the water, is used to simulate the buoyant behaviour of the snow. The proposed model shares many similarities with those used to investigate the impact of landslides into water bodies (e.g., see Fritz et al. 2003a, Fritz et al. 2003b), but it also clarifies the differences between the impact of an avalanche and that of a landslide: while a landslide typically reaches the bottom of the water body, because of its high constant density, in the present experiments a floating motion of the impacted mass has been observed.

The experimental setup that reproduces such model is composed of an inclined wooden chute reaching the bottom of a 3 m long prismatic flume filled with colored water, with transparent glass sidewalls. Both chute and flume are 0.11 m wide. The used still water depth is 0.14 m. A mass composed of granular expanded clay, is released at the top of the slope, thus sliding down and reaching the flume, where it enters the water body and generates the wave. Such model represents a two-dimensional approximation of the problem and is represented in Figure 1, where also some fundamental parameters are reported.

The wave generation and its propagation are acquired using both an high-speed camera, placed in the proximity of the impact zone, and two low-speed cameras, placed along the flume. The image processing of the acquired video leads to the determination of the instantaneous water elevation.

The amplitude decay, the height decay and the local values of the celerity of the leading wave are evaluated from the acquired elevation, using a proper space-time domain that allows to discard from the subsequent

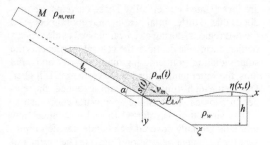

Figure 1. Two-dimensional physical model for an avalanche impacting into water.

analysis the impacting particles, their splashes and the reflection of the wave at the end of the flume.

Four experiments are analyzed, in which the variation of the mass and of the slope length also affects the velocity at the impact. The dimensionless form of the acquired data, i.e the amplitude decay, the height decay and the celerity of the wave, are compared with the corresponding literature predictive functions for impulse waves generated by landslides (Heller and Hager 2010). Results suggest that 1) the wave is strongly nonlinear in the vicinity of the impact and 2) the existence of a "wave formation zone", where the characteristics of the wave cannot be predicted with the common approximated theories.

REFERENCES

Fritz, H. M., W. H. Hager, & H.-E. Minor (2003a). Landslide generated impulse waves. 1. instantaneous flow fields. *Experiments in Fluids* 35(6), 505–519.

Fritz, H. M., W. H. Hager, & H.-E. Minor (2003b). Landslide generated impulse waves. 2. hydrodynamic impact craters. *Experiments in Fluids* 35(6), 520–532.

Heller, V. & W. H. Hager (2010). Impulse product parameter in landslide generated impulse waves. *Journal of waterway, port, coastal, and ocean engineering* 136, 145–155.

Two dimensional Lattice Boltzmann numerical simulation of a buoyant jet

A. Montessori, P. Prestininzi & M. La Rocca
Dipartimento di Ingegneria, Università degli Studi Roma Tre, Roma, Italy

D. Malcangio & M. Mossa
Dipartimento di Ingegneria Civile, Ambientale, del Territorio, Edile e di Chimica, Politecnico di Bari, Bari, Italy

ABSTRACT

Since its introduction (Succi, 2001), the Lattice Boltzmann Method (LBM) has obtained increasing attention in the field of Computational Fluid Dynamics, due to its intrinsic simplicity and ability even in dealing with rather complex flows. This work is aimed at assessing the ability of the two dimensional version of the LBM in simulating a buoyant jet. The latter is a saline jet (discharge Q_j, density ρ, diameter d) entering a uniform flow (discharge q, density ρ_0, height H). The inlet velocity of the jet is perpendicular to the velocity of the flow. The jet reaches the height z_{max} within the cross flow (Fig. 1).

The adopted two-dimensional Lattice Boltzmann formulation is equivalent to the Navier-Stokes equation with a Boussinesque gravity force term. Experiments on turbulent negatively buoyant jets into cross flow have been utilized to assess the validity of the numerical simulations.

A conductivity probe for high resolution measurement of turbulent density was utilized to measure the mixing of the salt water into the cross current, whereas the Acoustic Doppler Velocimeter (ADV) system was used for the measurement of the instantaneous three-flow velocity components. The comparison of the two dimensional numerical results with the experimental results shows that the former are able to capture the general appearance of the flow from a qualitative point of view. Moreover, a fair agreement is observed on macroscopic quantities (jet's maximum heigt, salinity at given points, etc.), defined in Gungor and Roberts (2009) and characterising the jet.

Differences arise as, due to the opening of the jet in the spanwise direction, three dimensional effects become important. It seems that the two dimensional simulations are characterised by a reduced ability in dissipating turbulent structures.

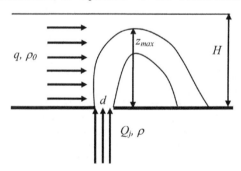

Figure 1. Sketch of the saline jet entering a cross flow.

REFERENCES

Gungor, E. and Roberts P.J.W., 2009. Experimental studies on vertical dense jets in a flowing current. Journal of Hydraulic Engineering 135 (11): 935–948.

Succi, S., 2001. The lattice Boltzmann equation: for fluid dynamics and beyond. Oxford: Oxford university press.

Interfacial instabilities of gravity currents in the presence of surface waves

L.M. Stancanelli, R.E. Musumeci & E. Foti
Department of Civil Engineering and Architecture, University of Catania, Catania, Italy

ABSTRACT

Gravity currents are also known as density currents, are produced by the density difference between two or more fluids. Such currents are quite common both in natural and man-made systems. Good examples are saline intrusion, outflows of industrial plants, oil spills in the ocean, turbid water intrusion in a lake, suspended sediment plume in a river or in coastal environment. The field investigation of the gravity currents is usually very difficult due to their complicated and unexpected occurrence characteristic (An et al. 2012). Notwithstanding the fact that the discharge of fresh or brackish water in the sea is frequent, the effect of the wave motion on the propagation in coastal regions of the salt-brackish wedge has not been systematically investigated yet (Robinson et al. 2013). The aim of the research presented here is to investigate Boussinesq gravity current by evaluating the influence of regular surface waves on the gravity currents dynamics. In particular, specifics objectives are the assessment of the front velocity changes due to the wave field and the study of mixing at the interface in such a complex situation. In order to cope with these problems both an experimental and a numerical approach have been adopted. In both cases the lock exchange schematization has been used. In particular, the numerical calculations reproduced the same conditions of the experimental set up. The commercial CDF model FLOW-3D was selected as the numerical model. A wide range of experiments have been performed considering different reduced gravity values (range of g' 0.04–0.15 m/s^2) and different wave conditions (i.e. wave height range 1.5–4.3 cm, wave period range 0.7–1.3 s). Preliminary analysis on experimental results in the absence of waves have been performed in order to validate the adopted measurement techniques and to use the results as a reference for the experiments carried out in the presence of waves. When analyzing the influence of regular waves on the gravity current dynamics, a strong interaction between the two flows has been observed and evaluated in terms of the dimension and of the velocity of the front of the heavier gravity current. The presence of the waves significantly modifies (up to 30%) the dynamics of the front propagation by inducing a general reduction of the front velocity depending on the value of the reduced gravity. The presence of the waves causes: i) the dynamics of the flow becoming pulsating with the same period

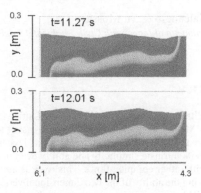

Figure 1. CFD simulation of gravity current $g' = 0.10$ m/s^2 and in the presence of surface regular waves ($H_w = 2.86$ m, $T_w = 0.84$ s, $L = 0.95$ m).

of the superimposed surface waves; ii) damping of the instabilities usually observed at the boundary between the denser and the lighter fluid. The numerical modeling carried out adopting the FLOW-3D model provided more information about the variation of salinity density at the interface and the influence of the wave motion (see Figure 1). In particular, the numerical simulation allows to increase the observational time, as in the experimental model only a 1 m long measuring area was available. The results of the present experimental campaign seem to confirm some of the results of previous works (Huppert and Simpson 1980, Robinson et al. 2013, Ng and Fu 2002). Future work is deemed necessary in order to extend the range of the numerical simulations, for example by varying the reduced gravity and the wave parameters and by also testing more complex turbulence model (i.e. LES method).

REFERENCES

An, S., S. K. Venayagamoorthy, & P.Y. Julien (2012). Numerical simulation of particle-driven gravity currents. *Environ Fluid Mech 12*, 495–513.

Huppert, H. E. & J. E. Simpson (1980). The slumping of gravity currents. *Journal of Fluid Mechanics 99*(04), 785–799.

Ng, C.-O. & S.-C. Fu (2002). On the propagation of a two-dimensional viscous density current under surface waves. *Physics of Fluids (1994-present) 14*(3), 970–984.

Robinson, T., I. Eames, & R. Simons (2013). Dense gravity currents moving beneath progressive free-surface water waves. *Journal of Fluid Mechanics 725*, 588–610.

Unraveling salt fluxes: A tool to determine flux components and dispersion rates from 3D models

W.M. Kranenburg, T. van der Kaaij, H. van den Boogaard & R.E. Uittenbogaard
Deltares, Delft, The Netherlands

Y.M. Dijkstra
Delft University of Technology, Delft, The Netherlands

ABSTRACT

In many delta's of the world salinity intrusion imposes limits to fresh water availability. Where globally the need for fresh water is increasing, at many places also salinity intrusion is expected to increase due to climate change through its effect on river discharge characteristics, sea levels and storm climates (WWAP 2015). In manipulated delta's like e.g. the Rhine-Meuse delta, salt intrusion is impacted by human activities as well. Salt intrusion is not only a relevant, but also a complex topic, due to the presence of a range of 3D mechanisms, and strong variations in space and time (Fischer et al. 1979, Geyer & McCready 2014). The relevance and complexity call for a proper system understanding (diagnosis), but also for means to predict and to explore salinity intrusion prevention measures based upon that (prognosis).

For both system understanding and prediction of intrusion in science and engineering practice, 3D numerical models play an important role. However, interpretation of 3D model output in terms of processes is complex: it is hard to see what is actually determining your results. Next, 3D models are often too heavy to carry out large numbers of computations, e.g. for derivation of salt intrusion statistics for future scenario's. For this type of work, often 1D models are used. However, with their limited process description and usual calibration on present or historic conditions, the value of this type of models for scenario's outside the calibration condition range or changes within the system, is limited.

In this paper, we present an instrument to determine salt flux components and equivalent dispersion rates from 3D numerical model results. The decomposition tool, inspired by Fischer et al. (1979) and Lerczak et al. (2006), consists of a spatial and a temporal decomposition. The first decomposes velocity and salinity in cross-sectional averages and variations to quantify contributions of cross-sectional means and variations to the total mean salt flux. The second separates the effect of storm-events and tidal components by processing velocity and salinity signals with Godin filtering in combination with harmonic

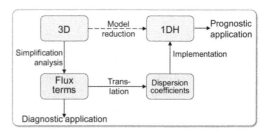

Figure 1. Schematic overview of connections between models and components of this study.

analysis. From the spatial decomposition, equivalent dispersion coefficients are computed and applied in a 3D-1D numerical model comparison, showing nearly complete reproduction.

We conclude that the salt flux decomposition instrument provides valuable support in interpretation of 3D-model results by giving insight in the relative importance of various physical mechanisms. The determination of equivalent dispersion provides means to verify 1D-models and to incorporate characteristics and qualities of 3D models into these faster 1D models. The tool is available and in principle applicable in combination with any 3D-numerical model.

REFERENCES

Fischer H.B., List, E.J., Koh, R.C.Y., Imberger, J. & Brooks, N.H. 1979. *Mixing in inland and coastal waters*. New York Academic Press.

Geyer, R.W. & MacCready, P. 2014. The estuarine circulation. *Ann. Rev. Fluid Mech.* 46: 175–197.

Lerczak J.A., Geyer, W.R. & Chant, R.J. 2006. Mechanisms driving the time-dependent salt flux in a partially stratified estuary. *Journal of Physical Oceanography*. 36: 2296–2311.

WWAP (United Nations World Water Assessment Programme). 2015. *The United Nations World Water Development Report 2015: Water for a Sustainable World*. Paris, UNESCO.

Density currents flowing up a slope

L. Ottolenghi & C. Adduce
Department of Engineering, Roma Tre University, Rome, Italy

R. Inghilesi
Institute for Environmental Protection and Research, Italy

V. Armenio
Department of Engineering and Architecture, University of Trieste, Trieste, Italy

F. Roman
Iefluids, Trieste, Italy

ABSTRACT

Gravity currents are flows in which the predominantly horizontal buoyancy gradient causes the propagation of a dense fluid into another at lower density (Simpson 1997). Gradients in the temperature or in the concentration fields generate the difference in density, and this circumstance widely occurs in nature both in the atmosphere and in the oceans (sea breeze fronts and oceanic overflows).

The lock-exchange technique is a common practice applied in the laboratory in order to generate gravity currents by the sudden release of a fixed volume of dense fluid into another ambient one of a lower density. This technique is based on the sudden interaction of two fluids in a tank divided by a vertical barrier. Ambient fluid at density ρ_0 fills the volume on the right-hand side of the barrier; dense fluid at density $\rho_1 > \rho_0$ fills the volume on the left-hand side. When the barrier is quickly removed, a gravity current forms and the dense fluid starts spreading on bottom of the tank under the ambient fluid, moving away from the lock region.

The lock-exchange configuration is here numerically reproduced in order to investigate the dynamics of unsteady gravity currents spreading along an upsloping bottom by Large Eddy Simulations (LES). The inclination of the bottom boundary, θ is varied in order to investigate the effect of an up-sloping bottom on the flow dynamics.

During the propagation of the gravity current, different flow regimes are generally identified (Rottman and Simpson 1983): a slumping phase followed by a self-similar phase. It is observed that during the slumping phase the presence of an upsloping bottom doesn't affect the front propagation of the gravity current, and the current flows at a constant value of velocity. On the contrary, during the self-similar phase, the gravity current slows down and the velocity of the front position decreases more abruptly with the increase of θ.

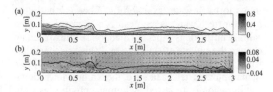

Figure 1. Dense current flowing up a slope: (a) density field; (b) velocity field. Isopycnals in the density field refer to $\rho^* = 2\%$, $\rho^* = 10\%$, $\rho^* = 20\%$ and $\rho^* = 50\%$.

A smoother behaviour of the current profile is also visible for high values of θ. Furthermore, during the self-similar regime the head of the gravity current becomes thinner in the upsloping cases. The development of a backward flow in the tail region of the gravity current is observed through the inspection of the density and velocity fields of the gravity current: dense fluid flows to the lock where an accumulation region forms.

Entrainment processes occurring between the ambient fluid and the dense current are also observed and investigated. During the propagation of a gravity current, mixing between the dense and the ambient fluids occurs. In fact, ambient fluid is entrained by the current, mixing with it and affecting the flow dynamics. It is found that the entrainment is affected by θ and, in particular, a decrease of the entrainment with the increase of the steepness of the bottom is observed.

REFERENCES

Rottman, J. & J. Simpson (1983). Gravity currents produced by instantaneous releases of a heavy fluid in a rectangular channel. *J. Fluid Mech.* 135, 95–110.

Simpson, J. E. (1997). *Gravity currents: In the environment and the laboratory*. Cambridge University Press.

Turbulent entrainment in a gravity current

M. Holzner
ETH Zurich, Switzerland

M. van Reeuwijk
Imperial College London, UK

H. Jonker
Delft University, The Netherlands

ABSTRACT

We revisit the classical entrainment experiments for gravity currents on inclined slopes [2]. We derive an entrainment relation that couples the entrainment rate E to the production of turbulence kinetic energy, the net effect of buoyancy and inner layer. Using direct numerical simulations (DNSs) we show that the net effect of inner layer processes on entrainment is very small and that buoyancy has an almost negligible effect on E. It is demonstrated that the dominant process causing entrainment is turbulence production due to shear. Second, we observe that for all simulations the eddy diffusivity and dissipation rate can be param-eterised using the turbulence kinetic energy and shear parameter. This information can be used to derive an entrainment law which is in good agreement with the DNS results. We discuss the potential reasons for why this result is significantly different from experiments and the classical entrainment law introduced by Ellison and Turner. Density currents are important in a variety of settings, ranging from oceanic to atmospheric flows, as well as hydraulic engineering applications (e.g. sediment laden gravity currents) [1–4]. A central aspect that controls the dynamics of density currents is the entrainment of external fluid into the turbulent flow, which is quantified by the "entrainment rate" E, i.e. the ratio between the entrainment velocity normal to the current w and the mean downstream velocity U, $E = w/U$. In this work we use DNSs of temporal gravity currents and we simulate in large domains and for long times to make sure that self similarity is reached. The figure shows E against Richardson number Ri for different simulations conducted. Our data shows an approximately linear decrease to zero for $Ri < 0.25$. The trend is very different from the one of [2], in particular the observed range in Ri is much smaller.

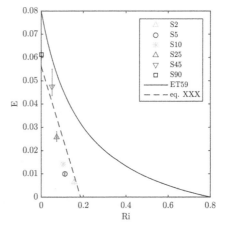

Figure 1. Entrainment rate as a function of Ri from DNS (symbols), showing that the observed range of Ri is much smaller than the Ellison-Turner relation (continuous line).

REFERENCES

[1] Cenedese, C. and Adduce, C., 2008. Mixing in a density-driven current flowing down a slope in a rotating fluid. J. Fluid Mech. 604, 369–388.

[2] Ellison, T. and Turner J., 1959. Turbulent entrainment in stratified flows. J. Fluid Mech. 6, 423–448.

[3] Krug, D. et al., 2015 The turbulent/non-turbulent interface in an inclined dense gravity current. J. Fluid Mech. 765, 303–324.

[4] Wells, M. et al., 2010. The relationship between flux coefficient and entrainment ratio in density currents. J. Phys. Oceanogr. 40, 2713–2727.

Remote sensing and coastal morphodynamic modelling: A review of current approaches and future perspectives

Mario Benincasa
Department of Engineering, Università Roma Tre, Roma, Italy

Federico Falcini
Institute of Atmospheric Sciences and Climate, CNR, Italy

Claudia Adduce
Department of Engineering, Università Roma Tre, Roma, Italy

Rosalia Santoleri
Institute of Atmospheric Sciences and Climate, CNR, Italy

ABSTRACT

We want to investigate the use of remotely sensed data in shoreline morphodynamic modelling. For this aim, we present a review of current approaches for Total Suspended Matter (TSM) retrieval from satellite data in coastal waters and outline some perspectives for the definition of an operative, satellite-based, alongshore sediment flux.

Coastal morphodynamic is generally modelled by means of the Exner equation, which describes the conservation of mass that relates the shoreline profile η to the divergence of the sediment flux Q_s

$$\frac{\partial \eta(x,t)}{\partial t} = -\frac{1}{D}\frac{\partial Q_s(x,t)}{\partial x}$$

The evaluation of the sediment flux ($Q_s(x,t)$) is usually a challenging task, subject to heavy assumptions and approximations. We here explore the possibility to evaluate the sediment flux from a remote sensing approach. We aim at combining a satellite-inferred sediment concentration together with a water velocity field, as obtained by e.g. coastal oceanographic models. With such a combination we can obtain a sediment flux and we can feed this sediment flux in a coastal morphodynamic model.

However, coastal, turbid and optically complex waters have traditionally posed challenges to remote sensing and space-borne techniques for a variety of reasons: the complexity and variability of their radiative processes on one hand; spatial, temporal and spectral resolution issues on the other. Nevertheless different sensors and different techniques have been used in order to retrieve the main biogeochemical characteristics of these waters.

A large variety of different methods, empirical or semi-analytical, are used by the scientific community in order to get the most suitable algorithm for the TSM (Total Suspended Matter) retrieval, and within the TSM, to discriminate the sediment (i.e. the mineral part of the TSM), and within the sediment, the geomorphologically active portion. Thus we outline the issues and the investigations of the optical behaviour of the different particles based on their composition, size and vertical distribution. Envisaged developments include multi sensor approaches, new algorithms, coupled sea-atmosphere radiative transfer models. We here discuss the best strategy in order to achieve the most suitable, regional TSM product for coastal geomorphologic applications and thus to pair it with coastal water velocity fields, allowing for a satellite-based definition of alongshore sediment transport. We also discuss spatial, temporal, and spectral characteristics of different sensors and novel algorithms that might combine these properties together.

A comparison of simple buoyant jet models with CFD analysis of overflow dredging plumes

B. Decrop
Dredging and Offshore Department, IMDC, Antwerp, Belgium

T. De Mulder
Department of Civil Engineering, Ghent University, Ghent, Belgium

E. Toorman
Department of Civil Engineering, KULeuven, Leuven, Belgium

ABSTRACT

Trailing Suction Hopper Dredgers release excess water with a varying flow rate and with variable fine sediment content (Figure 1). The released water-sediment-air mixture causes a negatively-buoyant sediment jet underneath the dredging vessel, which is also influenced by air bubbles entrained into the overflow shaft. The stage in which the plume is driven by buoyancy is called the near-field. Further from the vessel, the sediment concentration has reduced and the plume is no longer influenced by air bubbles. At this stage the plume drifts with the ambient current, this stage is called the far-field.

In the recent past, the near-field dispersion of overflow dredging plumes was determined using simple integral solutions or Lagrangian models of the buoyant jet in cross flow. In reality, these negatively-buoyant sediment plumes are interacting with the flow around the hull of the vessel, air bubbles and the propellers. If these interactions are not taken into account for the near-field modelling, the source terms for far-field simulations of the environmental impact of turbidity are inaccurate. By consequence, the predictions of the environmental impact of the generated turbidity might not be accurate enough to avoid adverse effects later on in the project phase.

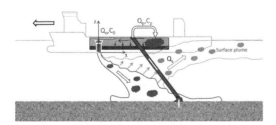

Figure 1. Situation sketch of a Trailing Suction Hopper Dredger working using the overflow shaft.

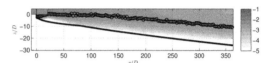

Figure 2. Vertical slice along the axis of a plume, with the CFD sediment concentration plotted in grey-scale ($\log(C/C_0)$), the CFD centerline in round markers and the theoretical solution in full line. Boundary conditions are $C_0 = 20$ g/l, $H = 39$ m, $W_0 = 3.2$ m/s, ship speed-through-water $U_0 = 1.5$ m/s and $D = 1.1$ m. The overflow was located at 20 m from the stern.

In a CFD analysis in Ansys Fluent, it is investigated how these complex interactions take place and how they can be included in near-field dredging plume simulations. The CFD analysis reveals that the simple models can be relatively accurate in some cases, but that large deviations exist for most real-life situations (like for example in figure 2).

An analysis of a large number of CFD cases shows that only exceptional cases of plumes without air bubbles, with strong buoyancy, weak crossflow and an overflow shaft located far from the stern, can be reproduced accurately by integral laws or simplified models. In all other cases, the simplified models predict plumes with a centerline significantly lower compared to the realistic CFD simulations. Moreover, from an environmental point of view, the surface plumes generated by air bubbles and propellers are most important. They cannot be reproduced by the commonly used simplified models while they are simulated correctly in the CFD model.

The CFD model results can be used either directly or indirectly by fitting a parameter model to the CFDdataset. Both options lead to more accurate sediment source terms for far-field dispersion models and by consequence more accurate environmental impact assessments of dredging plumes.

*Special session: Management of hydraulic systems
by means of fuzzy logic*

Assessment of annual hydrological drought based on fuzzy estimators

M. Spiliotis, P. Angelidis & B. Papadopoulos
Department of Civil Engineering, Democritus University of Thrace, Xanthi, Greece

ABSTRACT

Even if a variety of indices for assessing hydrological drought have been devised, which, in general, are data demanding and computationally intensive, rather simple and effective indices such as the Standardised Precipitation Index (SPI) have been widely used for hydrological droughts.

In this article, we propose a fuzzy version of the annual Streamflow Drought Index (SDI) which is an index analogous to SPI based on the cumulative streamflow volume (Nalbantis and Tsakiris, 2009). This drought index refers to the hydrological drought. The conventional SDI index is based on the ratio of the difference between the current cumulative volume and its mean value, to the standard deviation of the sample.

The mean and the standard deviation of annual cumulative streamflow volumes are estimated as fuzzy numbers with respect to the data. This is achieved by using the fuzzy estimator. By using the fuzzy estimators we can exploit all the statistical information for a selected level of confidence. In fact, fuzzy estimators can be characterized as a hybrid approach.

Thus, we can exploit the extension principle of fuzzy sets which enables us to define arithmetic operations with fuzzy sets.

The proposed method was applied on annual cumulative discharge of the transboundary Evros River. Hence, the monthly discharges of the transboundary Evros River at the bridge of Pythio (border of Greece and Turkey, downstream of the town of Edirne) are examined (Angelidis et al., 2010).

In brief, the proposed methodology consists of the following steps:

1. Based on the annual cumulative volume of streamflow, we determine the mean and the standard deviation of the sample.
2. Based on the fuzzy estimators (Chrysafis and Papadopoulos, 2008), the α-cuts of the mean and the standard deviation are determined for a selected degree of confidence, $1-\gamma$.

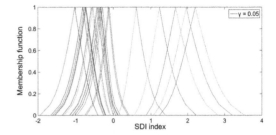

Figure 1. Historical sequences of the calculated fuzzy annual SDI for $\gamma = 0.05$.

3. For several α-cuts we determine the difference between the current annual cumulative volume of streamflow and its mean value, as it is estimated with the aid of fuzzy estimators.
4. For the same α-cuts with the previous step, and by using the concept of fuzzy division we calculate the value of the SDI index.

The outcomes are presented in Figure 1 An interesting point of view is that the fuzziness of the SDI increases for extremely wet or dry years. Another interesting point of view is that the fuzziness increases also regarding the large values of confidence interval ($1-\gamma$).

REFERENCES

Angelidis, P., Kotsikas, M., Kotsovinos, N., 2010. Management of Upstream Dams and Flood Protection of the Transboundary River Evros/Maritza. *Water Resources Management*, 24: 2467–2484.

Chrysafis, K.A., and Papadopoulos, B. K., 2008. On theoretical pricing of options with fuzzy estimators. *Journal of Computational and Applied Mathematics* 223: 552–566.

Nalbantis, I., Tsakiris, G., 2009. Assessment of Hydrological Drought Revisited. *Water Resources Management* 23(5): 881–897.

Fuzzy regression analysis between sediment transport rates and stream discharge in the case of two basins in northeastern Greece

M. Spiliotis
Department of Civil Engineering, Democritus University of Thrace, Xanthi, Greece

V. Kitsikoudis
Department of Civil Engineering, Istanbul Technical University, Istanbul, Turkey

V. Hrissanthou
Department of Civil Engineering, Democritus University of Thrace, Xanthi, Greece

ABSTRACT

Systematic measurements of sediment transport rates and water discharge were conducted in two basins, in northeastern Greece. Separate measurements of bed load transport and suspended load transport were performed in these basins located near Xanthi (Thrace, northeastern Greece): Kosynthos River basin with an area of about 237 km² and Kimmeria Torrent basin with an area of about 35 km². Measured data of rainfall depth, rainfall duration, water discharge and sediment transport for the outlets of the above basins were available.

In this study, relationships between sediment transport rates and water discharge are presented, based on nonlinear fuzzy regression due to the fact that there is insufficient and no absolutely reliable data. Thus, two curves were studied regarding the basins of Kosynthos River and Kimmeria Torrent, respectively: (i) the bed load transport rate, m_G, versus the water discharge, Q, (Fig. 1) and (ii) the suspended load transport rate versus the rainfall intensity and the water discharge (Fig. 2). The selection of the fuzzy curves is proposed based on the aims to minimize the total fuzziness and to avoid any overfitting behaviour.

It must be clarified that, according to the fuzzy regression model of Tanaka (1987), all the data must

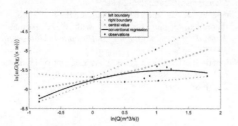

Figure 1. Graphical representation of logarithmic nonlinear (quadratic polynomial) fuzzy regression for the bed load transport rate m_G [kg/(sm)] with respect to the water discharge Q (m³/s), for Kosynthos River. The black curve represents the conventional regression.

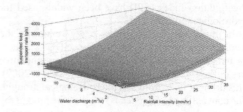

Figure 2. Graphical representation of the suspended load transport rate with respect to the rainfall intensity and water discharge, based on the nonlinear (quadratic polynomial) fuzzy regression, for Kimmeria Torrent.

be included in the produced fuzzy band. The fuzzy regression problem based on Tanaka (1987) method concludes to an constrained optimization problem (Kitsikoudis et al., 2016).

The first criterion of the successful implementation of the proposed method is a rational value of the sum of the fuzzy band with respect to all data. However, a complex relation could lead to curves with overfitting behaviour as in case of cubic fuzzy polynomial, especially if there is no enough data.

Another interesting point of view is that, if only one independent variable is taken into account, regarding the estimate of the suspended load transport rate, the objective function, that expresses the total fuzziness, increases significantly.

Finally, regarding the bed load transport rate, it can successfully be described either by using a quadratic polynomial fuzzy regression, or by using a logarithmic polynomial fuzzy regression, with respect to water discharge.

REFERENCES

Kitsikoudis, V., Spiliotis, M. and Hrissanthou, V., 2016. Fuzzy regression analysis for sediment incipient motion under turbulent flow conditions. *Environmental Processes*, in press (doi:10.1007/s40710-016-0154-2).

Tanaka, H., 1987. Fuzzy data analysis by possibilistic linear models. *Fuzzy Sets and Systems*, 24: 363–375.

Rainfall data regression model using fuzzy set logic

Christos Tzimopoulos & Chris Evangelides
Aristotle University of Thessaloniki, Greece

Basilios Papadopoulos
Democritus University of Thrace, Xanthi, Greece

ABSTRACT

Rainfall measurement models have been extensively used in the design process of water resource projects such as hydrological prediction, spillway design, climatic change studies, rainfall and runoff correlation etc. Rainfall measurements in a specific area are commonly displayed in the form of time series where recorded values can be either continuous or discrete. In many instances, there is a correlation between rainfall time series that belong to different stations and comprise measurements with differing range. E.g. there is an available time series of 15 years for station A and a time series of 30 years for station B. Due to the correlation between them, we can fill in the missing values, in order to extend the shorter time series. Correlation analysis is used to depict the relation between the dependent variable (usually the meteorological station with the shortest data-recording time span) and the independent variables (neighbouring stations with long recording time span). In classical linear regression, the difference between measurement values and estimated values, is a random variable with normal distribution and is considered to be caused by measurement errors. According to this, classical regression is considered to be probabilistic and has many uses but can be rendered problematic if the data set is small, if it's hard to prove that error distribution is normal, if there is fuzziness between dependent and independent variables or if linearity acceptance is not proper.

Nowadays, new regression models have been introduced by Tanaka et al (1982), Tanaka (1987), Redden and Woodall (1996), Savic and Pedrycz (1991), etc, based on fuzzy logic. In fuzzy regression, the difference between measurement values and estimated values is attributed to the inherent fuzziness of the system as well as to the fuzziness of input and output data. In contrast with classical regression analysis, fuzzy regression analysis uses fuzzy functions for the regression factors. The above problem usually meets one of the three cases: a) crisp input values x_{ij} and crisp output values y_j b) crisp input values x_{ij} and fuzzy output values $\tilde{y}_i$ c) fuzzy input values $\tilde{x}_{ij}$ and fuzzy output values $\tilde{y}_i$. In all of these cases, estimated values $\tilde{Y}_i$ are fuzzy. The adjustment of a fuzzy regression model can be achieved through two general methods: a) the possibilistic model and b) the least squares model. It is to point out that possibilistic parameters in the models are non interactive, i.e. the joint possibilistic distribution of parameters is defined by minimum operators.

In this article, quadratic membership functions, as defined by Celmiņš (1987) and Tanaka and Ishibuchi (1991), are considered to propose a method of interactive fuzzy parameters in possibilistic linear hydrological systems, located in the region of Central Macedonia (Northern Greece), and the method can be reduced to linear programming.

Optimal spatial allocation of groundwater under fuzzy hydraulic parameters

E. Sidiropoulos
Aristotle University, Thessaloniki, Greece

ABSTRACT

The present paper deals with a combined problem of water extraction and water allocation. The extraction of water involves groundwater pumping and aquifer exploitation, while the allocation introduces a spatial character into the whole process. The spatial aspect dominates and brings the problem to the area of spatial optimization. Related and sometimes more general problems of this category concern resource allocation and land use design. A genetic algorithm geared toward spatial problems has been presented by the author and utilized for related applications (e.g. Sidiropoulos and Fotakis, 2009 and 2011). This algorithm, called the operational genetic algorithm (OGA) is based on cellular concepts for local selection of attributes and for global emerging characteristics.

The need for adding fuzzy aspects to the treatment of the problems just mentioned is self-evident. Indeed, a lot of fuzzy analysis has been done in the area of groundwater management, although the subject is by no means yet closed. A lot less work can be found in the literature on fuzzy spatial optimization. Even less exists on combined problems, like the one presented here.

Indicatively, regarding groundwater, Woldt et al (1995) considered physical parameter uncertainty in modeling and prediction. In fuzzy groundwater management, a paper by Bogardi appeared as early as 1983, but most of the related work is to be found in later years. Guan and Aral (2004) considered hydraulic parameter uncertainty for the pump and treat remediation problem. They use fuzzy averages for the computation of their objective functions.

Fuzzy spatial optimization is only a recent subject of research, as indicated by Zhou et al (2015).

The present paper is part of an effort to lend fuzzy features to the above mentioned OGA, especially in the context of groundwater distribution. A simple physical aquifer model underlies the optimization process, so as to stress those aspects that concern the combination of fuzziness with optimization.

The hydraulic conductivity of the aquifer is represented as a fuzzy number. The same holds for the water needs of the various locations of the area under study. Optimal distributions of the water sources are determined by taking into account fuzziness in the hydraulic conductivity and in the water needs both separately or in combination. Also, the objective function consists either of the pumping cost or of the transport cost or of their sum, thus generating separate cases to be considered. The results are given in the form of fuzzy numbers, corresponding to the input fuzzy numbers for the hydraulic conductivity an of the water needs, by using α-cuts combined with the above genetic algorithm.

REFERENCES

Guan J. and M. M. Aral, 2004. "Optimal design of groundwater remediation systems using fuzzy set theory," Water Resources Research, vol. 40, no. 1, Article ID W01518, 20 pages.

Sidiropoulos E. and Fotakis, D., 2009. Cell-based genetic algorithm and simulated annealing for spatial groundwater allocation. WSEAS Transactions on Environment and Development, vol 5(4), pp. 351–360.

Sidiropoulos E. and Fotakis, D. 2011. Spatial optimization and resource allocation in a cellular automata framework in Cellular Automata: Simplicity behind Complexity, Salcido, A., editor, INTECH Scientific Publishing, Ch 4, pp. 67–87.

Woldt, W., Dou, C., Bogardi, I. and Dahab, M. 1995. Using fuzzy set methods to consider parameter imprecision in groundwater flow models. Models for Assessing and Monitoring Groundwater Quality (Proceedings of a Boulder Symposium, July 1995). IAHS Publ. no. 227, 1995.

Zhou M., Tan S., Tao L., Zhu X. and Akhmat G. 2015. An interval fuzzy land-use allocation model (IFLAM) for Beijing in association with environmental and ecological consideration under uncertainty. Quality and Quantity, vol. 49 (6), pp. 2269–2290.

Fuzzy logic uses on eutrophication and water quality predictions

G.K. Ellina & I. Kagkalou
Department of Civil Engineering, Democritus Univercity of Thrace, Xanthi, Greece

ABSTRACT

Aquatic Ecosystems and especially lakes are facing a vast number of human originated problems which most of the times appear to complicit in a great degree affecting ecosystems variously. Thus, the "response" to specific management practices is often non-linear while large lag time is noted. The interpretation and prediction of physical, chemical and biological functions of evolved ecosystems, is studied until today using widely available empirical and dynamic models and multi-criteria analysis methods.

The theory of fuzzy logic now applied in several research studies relevant to the assessment of water quality and trophic state of water bodies (Kung et al., Lu et al., Lu & Lo, 2006; Chen & Mynett 2003, Kung H, Ying L, Liu YC, 1992). Lake Karla (Thessaly, Greece) is a new reservoir which regenerated after 50 years of desiccation. It is a developing ecosystem because the structural features (abiotic and biotic) are highly variable; hence their functions (biogeochemical cycles, metabolism etc.) have a large degree of uncertainty.

Among the new and promising analyzing "tools" is the Fuzzy logic (fuzzy logic models) which contributes to the investigation of complex and volatile organic systems.

In the first part of this work the investigation of eutrophication factors is performed on the highly variable system of Lake Karla with the help of fuzzy logic theory contributing to a better understanding of the mechanisms that determine the function. These factors are Water Temperature, Nitrate, Total Phosphorus and Secchi Depth, while the chlorophyll's concentration has already been sampled. The goal is to acquire additional knowledge for effective protection and management of this new ecosystem.

Furthermore, by using Fuzzy Logic Software are developing methods suitable for Water Quality pre-evaluation on in the studying area. The purpose is the uncertainty and fuzziness deduction towards the

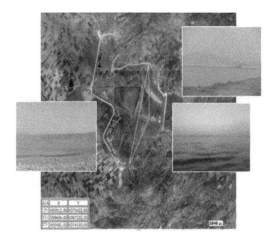

Figure 1. The study area lake Karla and sampling stations.

criteria that are used in decision making tools. More precisely, the factors being studied for the evaluation of water quality in Karla Lake are pH, Total Phosphorus and Nitrate. In this way, these indicators are classified to predict the Water Quality of Lake Karla.

REFERENCES

Bartsch A.F, Gakstatter J.H. (1978). Management decisions for lake systems on a survey of trophic status, limiting nutrients and nutrient loadings, in: American–Soviet. *Symposium on Use of Mathematical Models to Optimize Water Quality Management* (U.S. EPA Office of Research and Development, ERL, Gulf Breeze, FL, 1978) pp. 372–394, EPA-600/9-78-024.

Kung H., Ying L., Liu Y.C. (1992) A complementary tool to water quality index: fuzzy clustering analysis. *Water Resources.* Bull; 28(3):525–33.

Papadopoulos B.K., Sirpi M.A. (2004). Similarities and distances in fuzzy regression models, *Soft Computing*, 8 (8) 556–551.

Author index

Abadie, S. 70
Adam, N.J. 165
Adduce, C. 178, 180
Agarwal, A. 49
AghaKouchak, A. 3
Ahmadian, R. 18
Akay, H. 77
Al Zohbi, G. 52
Al-Qurashi, A. 143
Albentosa Hernández, E. 136
Alhasan, Z. 93
Altan-Sakarya, A.B. 108
Alvarado Montero, R. 166
Álvarez-Enjo, M. 155
Amacher, R. 73
Ancey, C. 174
Andrés-Doménech, I. 136
Angelidis, P. 185
Angelis, A.F. 27
Archambeau, P. 40, 83, 84, 92, 97, 102, 150, 158
Armanini, A. 90
Armenio, V. 178
Arrighi, C. 147, 152
Ata, R. 70
Aufleger, M. 112
Auguste, T. 109

Baduna Koçyiğit, M. 77
Bakar, A.A. 18
Balabanis, P. 4
Balmes, J. 146
Balzano, A. 62
Banda, M.S. 126
Basolo, V. 3
Bauwens, W. 86
Beaudouin, A. 88
Beersma, J. 140
Beevers, L. 153
Bellos, V. 125, 127
Benincasa, M. 180
Benoit, M. 70
Bercovitz, Y. 34
Bermudez, M. 149
Bialik, R.J. 33
Biniaş, J. 166
Blaise, S. 129
Blanckaert, J. 156
Bockelmann-Evans, B. 89
Bodeux, S. 169
Boelens, T. 67
Boettcher, H. 112

Bojkov, V. 94
Bolle, A. 71
Bombar, G. 29, 82
Boudreau, D. 3
Bouhlila, R. 17
Bousmar, D. 107, 109
Bozkuş, Z. 11
Brasiliano, L.N. 30
Brehin, F.G. 71
Bremer, F. 114
Brion, N. 19
Brisson, E. 141
Brocchini, M. 174
Brouyère, S. 169
Bruwier, M. 150
Bubalo, A. 100
Buiteveld, H. 140
Buldgen, L. 104
Bung, D.B. 14, 32
Buvat, C. 34

Calamak, M. 101
Caleffi, V. 123
Camus, P. 133
Carbonnel, V. 19
Carlier d'Odeigne, O. 118
Carraro, F. 123
Carrillo, J.M. 31, 95
Cassan, L. 8
Castanedo, S. 35
Castelli, F. 147, 152
Castillo, L.G. 31, 95
Castro, M. 95
Cea, L. 149, 155
Chamoglou, M. 20
Chen, M. 61, 86
Cheung, W. 3
Chicheportiche, J. 70
Christelis, V. 127
Chuquet, K. 159
Claeys, P. 19
Claeys, S. 38
Cleynen, O. 53, 54
Clous, L. 70
Coelho, P.S. 15
Cohen Liechti, T. 73
Collet, L. 153
Coquerez, S. 149
Crispino, G. 103

Dartus, D. 8
das Neves, L. 96

Dassargues, A. 169
David, L. 88
De Cesare, G. 73, 165
De Maeyer, J. 66
De Mulder, T. 67, 105, 110, 181
De Sutter, R. 65
de Ville, N. 28
Debenest, G. 9
Decrop, B. 181
Defina, A. 64
Dehant, V. 129
del Jesus, M. 133
Delandmeter, P. 129
Deleersnijder, E. 129
Delgado, R. 65
Deshpande, V. 87
Devaux, Y. 88
Devriese, P. 66
Dewals, B. 40, 83, 84, 92, 97, 102, 150, 158
Dezillie, N. 98
Dijkstra, Y.M. 177
Dinçer, A.E. 11
Diogo, P.A. 15
Dolinaj, D. 137
Doniec, A. 159
Đorđević, D. 121
Dornstädter, J. 98
Doucement, S. 52
Drogue, G. 158
Duchan, D. 93
Duma, D. 92
Dunstan, A. 69
Duru, A. 108
Duviella, E. 159

Egashira, S. 126
Eguiarte, A. 3
Eijgenraam, C.J.J. 68
El Azhari, M. 80
El kadi Abderrezzak, K. 97
Ellina, G.K. 189
Ellison, J. 55
Elskens, M. 19
Ennajem, L. 139
Eriş, E. 29
Erpicum, S. 40, 83, 84, 92, 97, 102, 150
Evangelides, C. 187

Falcini, F. 180
Falconer, R.A. 18

Farias, C.A.S. 30
Fazeres-Ferradosa, T. 96
Feldman, D.L. 3
Ferrand, M. 119
Ferreira, R.M.L. 10
Fichefet, T. 129
Filianoti, P. 168
Florescu, D. 47
Foti, E. 176
Fraga, I. 149, 155
Franca, M.J. 167

Gabl, R. 112
Gailler, A. 70
García, J.T. 31
García-Bartual, R. 136
Geiger, F. 48
Ghaïtanellis, A. 119
Giantsi, Th. 63
Gisonni, C. 103
Glas, M. 21
Gmeiner, P. 21
Goffin, L. 83
Golian, S. 134
Golzar, M. 128
Gonomy, N. 24, 28
Goodrich, K. 3
Goormans, T. 55, 156
Gourbesville, P. 120
Gourgue, O. 61
Grelier, B. 158
Gruwez, V. 71
Guillén-Ludeña, S. 13
Gutierrez, R.M. 36
Guymer, I. 128

Habersack, H. 21
Hajczak, A. 34
Hamzaoui-Azaza, F. 17
Hanert, E. 129
Hashemi, S.R. 138
Hassen, I. 17
Haut, B. 52
He, C. 25
Hébert, H. 70
Hegnauer, M. 140
Heidinger, P. 98
Hendrick, P. 51, 52
Hermosa, D. 95
Hidalgo, I.G. 27
Hidalgo, X. 95
Hoerner, S. 54
Hoitink, T. 129
Holzbecher, E. 143
Holzner, M. 179
Hosseinzadehtalaei, P. 141
Houston, D. 3
Hrissanthou, V. 186
Huang, H. 157

Ilic, S. 46
Ilinich, V.V. 160

Imbert, D. 70
Ingabire, J.S. 170
Inghilesi, R. 178

Jalili Ghazizadeh, M.R. 116
Jan Nesar, F. 138
Jin, Y. 145
Jodeau, M. 34
Joly, A. 70, 119
Jonker, H. 179
Joseph, A. 24
Jouini, M. 9
Journeau, C. 70
Julínek, T. 16

Kabiri-Samani, A. 115, 116, 117
Kagkalou, I. 20, 189
Karagiannis, N. 60
Karambas, T. 60
Karatekin, O. 129
Karpiński, M. 33
Kavaklı, Ü. 29
Kawuwa, A.S. 122
Kazolea, M. 70
Kemper, S. 146
Kentel, E. 72
Khan, A. 135
Khan, M. 135
Khashei Suiki, A. 138
Khellaf, M.C. 59
Kitsikoudis, V. 186
Klein, J. 12
Kløve, B. 23
Koca, K. 39
Koch, M. 26, 78
Kok, M. 68
Kokpinar, M.A. 108
Koppe, B. 161
Korkmaz, S. 79
Kortenhaus, A. 66
Koutitas, C. 60
Kranenburg, W.M. 177
Kumar, B. 87

La Rocca, M. 175
Lambrechts, J. 129
Lammersen, R. 140
Larina, T.D. 160
Launay, G. 106
Lavidas, G. 49
Le Bars, Y. 129
Le Gal, M. 7, 70
Le Roy, S. 70
Le Sourne, H. 104
Lebert, F. 34
Lee, S. 145
Legat, V. 129
Leroy, A. 119
Leščešen, I. 137
Lichtenberg, N. 53
Liedermann, M. 21

Lino, R.F. 15
Łoboda, A.M. 33
Lodomez, M. 102
Lopes, J.E.G. 27
Lopez, D. 13
López-Gallego, M. 50
López-Querol, S. 122
Lorke, A. 39
Losada, I.J. 133
Loudyi, D. 80, 139
Louis, S. 28
Lozano, J. 36
Luke, A. 3

Mabrouka, M. 81
Machado, E.C.M. 30
Mahesh, P. 87
Malcangio, D. 175
Manso, P. 167
Marcer, R. 70
Marchant, A. 128
Marence, M. 170
Marino, M.C. 85
Marion, A. 41
Marques, M.G. 111
Martin Medina, M. 70
Martzikos, N. 63
Masbernat, L. 9
Matthew, R.A. 3
Maude, F. 143
Mazzanti, B. 152
Meert, P. 142, 151
Merkel, U.H. 99
Mhashhash, A. 89
Michaux, B. 56
Mignot, E. 13, 106
Minns, A.W. 68
Miozzo, C. 41
Moftakhari, H.R. 3
Mohammadighavam, S. 23
Monbaliu, J. 66
Montessori, A. 175
Morabito, A. 51
Moradi Kashkooli, S. 138
Moreta, P.M. 122
Morichon, D. 70
Morreale, G. 168
Mossa, M. 175
Moutzouris, C.I. 63
Murla Tuyls, D. 148
Mustafa, A. 150
Musumeci, R.E. 176

Naithani, J. 129
Naumann, U. 99
Nekooie, M.A. 115, 116, 117
Nelko, V. 66
Ngoc Duong, V. 120
Nichols, A. 37, 41
Nicolai, R.P. 68
Niemann, A. 22, 166

Nóbrega, J.D. 111
Noss, C. 39
Notelé, R. 55
Nouasse, H. 159
Ntegeka, V. 148

O'Keefe, L. 46
Oertel, M. 12, 14, 45, 114
Oliveto, G. 85
Orban, P. 169
Ortega, P. 95
Ottolenghi, L. 178
Oumeraci, H. 147

Palomar, P. 35
Pan, S. 89
Papacharalampous, G. 63
Papadopoulos, B. 185, 187
Paredes-Morales, M. 143
Parvaneh, A. 115, 116, 117
Pedreros, R. 70
Peeters, P. 156
Pelaprat, L. 34
Peltier, Y. 40
Penchev, P. 94
Penchev, V. 94
Perdigão, D. 50
Pereira, F. 86, 151
Pérez-Díaz, B. 35
Petazzi, A. 155
Pfister, M. 73, 74, 103
Pham Van, C. 129
Piffer, S. 144
Pinho-Ribeiro, J. 50
Piro, C. 56
Pirotton, M. 40, 83, 84, 92, 97, 102, 150, 158
Plancke, Y. 38, 91
Pons, K. 70
Popa, B. 47
Popa, F. 47
Postacchini, M. 174
Pourreza-Bilondi, M. 154
Prestininzi, P. 175
Prinos, P. 173
Prudhomme, C. 153
Przyborowski, Ł. 33
Puertas, J. 149, 155
Pujades, E. 169

Radzi, A.H. 69
Ramos, H.M. 167
Ramos, P.X. 110
Rauwoens, P. 66
Reichl, W. 88
Reis, M.T. 96
Remacle, J.-F. 129
Ricardo, A.M. 10
Ricchiuto, M. 70
Rico Cortés, M. 136
Riehme, J. 99

Rifai, I. 97
Rigo, P. 104
Říha, J. 16, 93
Rijke, R.J. 55
Ritsch, S. 112
Rivière, N. 13, 106
Rizzi, A. 144
Rocha, T. 27
Rodrigues, A.C. 15
Roelandts, O. 118
Roman, F. 178
Romdhane, H. 8
Rosa-Santos, P. 50
Rosatti, G. 124, 144
Rossi, G. 90
Rotimi, A.J. 122
Rousseau, M. 70
Rubinato, M. 37
Ruigar, H. 134
Rutschmann, B. 48
Rutschmann, P. 48

Saeed, S. 141
Şahin, A.N. 11
Salsón, S. 155
Sammartano, V. 168
Samora, I. 167
Sanders, B.F. 3
Santoleri, R. 180
Sassi, M. 129
Savary, C. 107, 109
Schleiss, A.J. 73, 74, 165, 167
Schlenkhoff, A. 146
Schramkowski, G. 67
Schubert, J.E. 3
Schulz, H.E. 111
Schuttelaars, H. 67
Schäfer, S. 48
Serrano, K. 3
Šeta, B. 100
Sidiropoulos, E. 188
Sidiropoulos, P. 20
Silin, N. 66
Silva Jacinto, R. 70
Silva, R. 65
Sinagra, M. 168
Sishah, B.B. 61
Smets, S. 55
Soares-Frazão, S. 24, 118, 129
Sonnenwald, F. 128
Soualmia, A. 8, 9
Sperna Weiland, F.C. 140
Spiliotis, M. 185, 186
Stancanelli, L.M. 176
Steimes, J. 51, 52
Stojnić, I. 121
Stovin, V. 128
Stratigaki, V. 66
Štrbac, D. 137
Suzuki, T. 66
Swartenbroekx, C. 107, 109

Tabari, H. 141
Tait, S.J. 41
Taks, B. 170
Taveira-Pinto, F. 50, 96
Taştan, K. 113
Teller, J. 150
Teschlade, D. 22
Thant, S. 38
Thomas, C. 129
Thomas, L. 88
Thévenin, D. 53, 54
Toorman, E. 66, 181
Torlak, M. 100
Touhami, H.E. 59
Tritthart, M. 21
Troch, P. 66
Tsakiri, M. 173
Tsakiris, G. 125, 127
Tucciarelli, T. 168
Türkkan, G.E. 79
Tzimopoulos, C. 187

Uittenbogaard, R.E. 177
Urošev, M. 137

Valdés, J.M. 36
Valero, D. 14, 32
Valiani, A. 123
Vallaeys, V. 129
Van den Boogaard, H. 140, 177
van der Kaaij, T. 177
van Griensven, A. 86
Van Hoestenberghe, T. 98
Van Lipzig, N. 141
Van Looveren, R. 156
Van Oyen, T. 66
Van Ransbeeck, N. 98
van Reeuwijk, M. 179
van Vuren, S. 68
Vanderveken, J. 55
Vanneste, D. 66
Vanthillo, R. 98
Ventroni, M. 62
Venugopal, V. 49
Verbanck, M.A. 19, 28
Vercruysse, J. 105, 110
Verelst, K. 105, 110, 156
Vermuyten, E. 142
Verschoore, T. 118
Verwaest, T. 66
Viero, D.P. 64
Vigueras-Rodríguez, A. 31
Vincent, D. 129
Violeau, D. 7, 34, 70, 97, 119
Vos, G. 91
Vouaillat, G. 106

Wan Mohtar, W.H.M. 69
Westhoff, M.C. 84
Willems, P. 141, 142, 148, 151
Wilson, N. 128
Wolanski, E. 129

Woldegiorgis, B.T. 86
Wolfs, V. 142, 148
Wortberg, T. 166
Wüthrich, D. 74

Yanmaz, A.M. 77, 101
Yannie, A.B. 69

Yavari, P. 134
Yavuz, C. 72

Zahra Samadi, S. 154
Zare, M. 26, 78
Zech, Y. 24, 118
Zehe, E. 84

Zhang, L. 25
Zijlema, M. 62
Zimmermann, N. 71
Zitti, G. 174
Zorzan, G. 109
Zorzi, N. 144
Zugliani, D. 124, 144